Springer Undergraduate Mathematics Series

T0225881

For other titles published in this series, go to
www.springer.com/series/3423

David F. Griffiths · Desmond J. Higham

Numerical Methods for Ordinary Differential Equations

Initial Value Problems

 Springer

David F. Griffiths
Mathematics Division
University of Dundee
Dundee
dfg@maths.dundee.ac.uk

Desmond J. Higham
Department of Mathematics and Statistics
University of Strathclyde
Glasgow
djh@maths.strath.ac.uk

Springer Undergraduate Mathematics Series ISSN 1615-2085
ISBN 978-0-85729-147-9   e-ISBN 978-0-85729-148-6
DOI 10.1007/978-0-85729-148-6
Springer London Dordrecht Heidelberg New York

British Library Cataloguing in Publication Data
A catalogue record for this book is available from the British Library

Library of Congress Control Number: 2010937859

Mathematics Subject Classification (2010): 65L05, 65L20, 65L06, 65L04, 65L07

Printed on acid-free paper

Springer is part of Springer Science+Business Media (www.springer.com)

To the Dundee Numerical Analysis Group past and present,
of which we are proud to have been a part.

Preface

Differential equations, which describe how quantities change across time or space, arise naturally in science and engineering, and indeed in almost every field of study where measurements can be taken. For this reason, students from a wide range of disciplines learn the fundamentals of calculus. They meet differentiation and its evil twin, integration, and they are taught how to solve some carefully chosen examples. These traditional pencil-and-paper techniques provide an excellent means to put across the underlying theory, but they have limited practical value. Most realistic mathematical models cannot be solved in this way; instead, they must be dealt with by computational methods that deliver approximate solutions.

Since the advent of digital computers in the mid 20th century, a vast amount of effort has been expended in designing, analysing and applying computational techniques for differential equations. The topic has reached such a level of importance that undergraduate students in mathematics, engineering, and physical sciences are typically exposed to one or more courses in the area. This book provides material for the typical first course—a short (20- to 30-hour) introduction to numerical methods for initial-value ordinary differential equations (ODEs). It is our belief that, in addition to exposing students to core ideas in numerical analysis, this type of course can also highlight the *usefulness* of tools from calculus and analysis, in the best traditions of applied mathematics.

As a prerequisite, we assume a level of background knowledge consistent with a standard first course in calculus (Taylor series, chain rule, $\mathcal{O}(h)$ notation, solving linear constant-coefficient ODEs). Some key results are summarized in Appendices B–D. For students wishing to brush up further on these topics, there are many textbooks on the market, including [12, 38, 64, 70], and a plethora of on-line material can be reached via any reputable search engine.

There are already several undergraduate-level textbooks available that cover numerical methods for initial-value ODEs, and many other general-purpose numerical analysis texts devote one or more chapters to this topic. However, we feel that there is a niche for a well-focused and elementary text that concentrates on mathematical issues without losing sight of the applied nature of the topic. This fits in with the general philosophy of the Springer Undergraduate Mathematics Series (SUMS), which aims to produce *practical* and *concise* texts for undergraduates in mathematics and the sciences worldwide.

Based on many years of experience in teaching calculus and numerical analysis to students in mathematics, engineering, and physical sciences, we have chosen to follow the tried-and-tested format of Definition/Theorem/Proof, omitting some of the more technical proofs. We believe that this type of structure allows students to appreciate how a useful theory can be built up in a sequence of logical steps. Within this formalization we have included a wealth of theoretical and computational examples that motivate and illustrate the ideas. The material is broken down into manageable chapters, which are intended to represent one or two hours of lecturing. In keeping with the style of a typical lecture, we are happy to repeat material (such as the specification of our initial-value ODE, or the general form of a linear multistep method) rather than frequently cross-reference between separate chapters. Each chapter ends with a set of exercises that serve both to fill in some details and allow students to test their understanding. We have used a starring system: one star ($\star$) for exercises with short/simple answers, moving up to three stars ($\star\star\star$) for longer/harder exercises. Outline solutions to all exercises are available to authorized instructors at the book's website, which is available via `http://www.springer.com`. This website also has links to useful resources and will host a list of corrections to the book (feel free to send us those).

To produce an elementary book like this, a number of tough decisions must be made about what to leave out. Our omissions can be grouped into two categories.

Theoretical We have glossed over the subject of existence and uniqueness of solutions for ODEs. Rigorous analysis is beyond the scope of this book, and very little understanding is imparted by simply stating without proof the usual global Lipschitz conditions (which fail to be satisfied by most realistic ODE models). Hence, our approach is always to assume that the ODE has smooth solutions. From a numerical analysis perspective, we have reluctantly shied away from a general-purpose Gronwall-style convergence analysis and instead study global error propagation only for linear, scalar constant-coefficient ODEs. Convergence results are then stated, without proof, more generally. This book has a strong emphasis on numerical stability and qualitative properties of numerical methods hence; implicit methods have a prominent role. However, we do not attempt to study general conditions under which implicit recurrences have unique solutions, and how they can be computed. Instead, we give simple examples and pose exercises that illustrate some of the issues involved. Finally, although it would be mathematically elegant to deal with systems of ODEs throughout the book, we have found that students are much more comfortable with scalar problems. Hence, where possible, we do method development and analysis on the scalar case, and then explain what, if anything, must be changed to accommodate systems. To minimize confusion, we reserve a bold mathematical font $(x, f, \dots)$ for vector-valued quantities.

Practical This book does not set programming exercises: any computations that are required can be done quickly with a calculator. We feel that this is in keeping with the style of SUMS books; also, with the extensive range of high-quality ODE software available in the public domain, it could be argued that there is little need for students to write their own low-level computer code. The main aim of this book is to give students an understanding of what goes on "under the hood" in scientific computing software, and to equip them with a feel for the strengths and limitations of numerical methods. However, we strongly encourage readers to copy the experiments in the book using whatever computational tools they have available, and we hope that this material will encourage students to take further courses with a more practical scientific computing flavour. We recommend the texts Ascher and Petzold [2] and Shampine et al. [62, 63] for accessible treatments of the practical side of ODE computations and references to state-of-the-art software. Most computations in this book were carried out in the MATLAB© environment [34, Chapter 12].

By keeping the content tightly focused, we were able to make space for some modern material that, in our opinion, deserves a higher profile outside the research literature. We have chosen four topics that (a) can be dealt with using fairly simple mathematical concepts, and (b) give an indication of the current open challenges in the area:

1. Nonlinear dynamics: spurious fixed points and period two solutions.

2. Modified equations: construction, analysis, and interpretation.

3. Geometric integration: linear and quadratic invariants, symplecticness.

4. Stochastic differential equations: Brownian motion, Euler–Maruyama, weak and strong convergence.

The field of numerical methods for initial-value ODEs is fortunate to be blessed with several high-quality, comprehensive research-level monographs.

Rather than pepper this book with references to the same classic texts, it seems more appropriate to state here that, for all general numerical ODE questions, the oracles are [6], [28] and [29]: that is,

J. C. Butcher, *Numerical Methods for Ordinary Differential Equations*, 2nd edition, Wiley, 2008,

E. Hairer, S. P. Nørsett and G. Wanner, *Solving Ordinary Differential Equations I: Nonstiff Problems*, 2nd edition, Springer, 1993,

E. Hairer and G. Wanner, *Solving Ordinary Differential Equations II: Stiff and Differential-Algebraic Problems*, 2nd edition, Springer, 1996.

To learn more about the topics touched on in Chapters 12–16, we recommend Stuart and Humphries [65] for numerical dynamics, Hairer et al. [26], Leimkuhler and Reich [45], and Sanz-Serna and Calvo [61] for geometric integration, and Kloeden and Platen [42] and Milstein and Tretyakov [52] for stochastic differential equations.

Acknowledgements We thank Niall Dodds, Christian Lubich and J. M. (Chus) Sanz-Serna for their careful reading of the manuscript. We are also grateful to the anonymous reviewers used by the publisher, who made many valuable suggestions.

We are particularly indebted to our former colleagues A. R. (Ron) Mitchell and J. D. (Jack) Lambert for their influence on us in all aspects of numerical differential equations.

It would also be remiss of us not to mention the patience and helpfulness of our editors, Karen Borthwick and Lauren Stoney, who did much to encourage us towards the finishing line.

Finally, we thank members of our families, Anne, Sarah, Freya, Philip, Louise, Oliver, Catherine, Theo, Sophie, and Lucas, for their love and support.

DFG, DJH
August 2010

Contents

1

ODEs—An Introduction

Mathematical models in a vast range of disciplines, from science and technology to sociology and business, describe how quantities *change*. This leads naturally to the language of ordinary differential equations (ODEs). Typically, we first encounter ODEs in basic calculus courses, and we see examples that can be solved with pencil-and-paper techniques. This way, we learn about ODEs that are linear (constant or variable coefficient), homogeneous or inhomogeneous, separable, etc. Other ODEs not belonging to one of these classes may also be solvable by special one-off tricks. However, what motivates this book is the fact that the overwhelming majority of ODEs do not have solutions that can be expressed in terms of simple functions. Just as there is no formula in terms of a and b for the integral

$$\int_a^b e^{-t^2}\,dt,$$

there is generally no way to solve ODEs exactly. For this reason, we must rely on *numerical methods* that produce approximations to the desired solutions. Since the advent of widespread digital computing in the 1960s, a great many theoretical and practical developments have been made in this area, and new ideas continue to emerge. This introductory book on numerical ODEs is therefore able to draw on a well-established body of knowledge and also hint at current challenges.

We begin by introducing some simple examples of ODEs and by motivating the need for numerical approximations.

D.F. Griffiths, D.J. Higham, *Numerical Methods for Ordinary Differential Equations*, Springer Undergraduate Mathematics Series, DOI 10.1007/978-0-85729-148-6_1, © Springer-Verlag London Limited 2010

Example 1.1

The ODE[1]

$$x'(t) = \sin(t) - x(t) \tag{1.1}$$

has a general solution given by the formula $x(t) = A\,e^{-t} + \frac{1}{2}\sin(t) - \frac{1}{2}\cos(t)$, where A is an arbitrary constant. No such formula is known for the equation

$$x'(t) = \sin(t) - 0.1x^3(t), \tag{1.2}$$

although its solutions, shown in Figure 1.1(right), have a remarkable similarity to those of equation (1.1) (shown on the left) for the same range of starting points (depicted by solid dots: •).   □

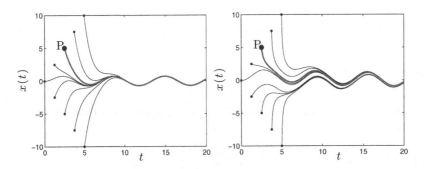

Fig. 1.1   Solution curves for the ODEs (1.1) and (1.2) in Example 1.1

Although this book concerns numerical methods, it is informative to see how simple ODEs like (1.1) can be solved by pencil-and-paper techniques. To determine the exact solutions of linear problems of the form

$$x'(t) = \lambda(t)x(t) + f(t) \tag{1.3}$$

we may first consider the homogeneous equation[2] that arises when we ignore terms not involving x or its derivatives. Solving $x'(t) = \lambda(t)x(t)$ gives the so-called *complementary function* $x(t) = Ag(t)$, where

$$g(t) = \exp\left(\int_0^t \lambda(s)\,ds\right)$$

---

[1]Here, and throughout, we shall use the notation $x'(t)$ to denote the derivative $\frac{d}{dt}x(t)$ of $x(t)$.

[2]A function $F(x)$ is said to be homogeneous (of degree one) if $F(\lambda x) = \lambda F(x)$. It implies that the equation $F(x) = 0$ remains unchanged if x is replaced by λx.

is the *integrating factor* and A is an arbitrary constant. The original problem can now be solved by a process known as *variation of constants*. A solution is sought in which the constant A is replaced by a function of t. That is, we seek a *particular solution* of the form $x(t) = a(t)g(t)$. Substituting this into (1.3) and simplifying the result gives

$$a'(t) = \frac{f(t)}{g(t)},$$

which allows $a(t)$ to be expressed as an integral. The general solution is obtained by adding this particular solution to the complementary function, so that

$$x(t) = Ag(t) + g(t) \int_0^t \frac{f(s)}{g(s)} \, ds. \tag{1.4}$$

This formula is often useful in the theoretical study of differential equations, and its analogue for sequences will be used to analyse numerical methods in later chapters; its utility for constructing solutions is limited by the need to evaluate integrals.

Example 1.2

The differential equation

$$x'(t) + \frac{2}{1 + t^4} x(t) = \sin(t)$$

has the impressive looking integrating factor

$$g(t) = \left\{ \frac{\exp\left[2\tan^{-1}(1 + \sqrt{2}t)\right] t^2 + \sqrt{2}t + 1}{\exp\left[2\tan^{-1}(1 - \sqrt{2}t)\right] t^2 - \sqrt{2}t + 1} \right\}^{1/(2\sqrt{2})},$$

which can be used to give the general solution

$$x(t) = Ag(t) + g(t) \int_0^t \frac{\sin(s)}{g(s)} \, ds.$$

The integral can be shown to exist for all $t \geq 0$, but it cannot be evaluated in closed form. Solutions from a variety of starting points are shown in Figure 1.2 (these were computed using a Runge–Kutta method of a type we will discuss later). This illustrates the fact that knowledge of an integrating factor may be useful in deducing properties of the solution but, to compute and draw graphs of trajectories, it may be much more efficient to solve the initial-value problem (IVP) using a suitable numerical method. □

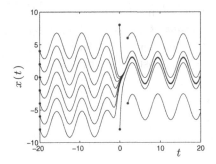

Fig. 1.2 Solution curves for the ODE in Example 1.2. The various starting values are indicated by solid dots ($\bullet$)

In this book we are concerned with first-order ODEs of the general form

$$x'(t) = f(t, x(t)) \tag{1.5}$$

in which $f(t, x)$ is a given function. For example, $f(t, x) = \sin(t) - x$ in (1.1), and in (1.2) we have $f(t, x) = \sin(t) - 0.1x^3$.

The solutions of equations such as (1.5) form a family of curves in the (t, x) plane—as illustrated in Figure 1.1—and our task is to determine just one curve from that family, namely the one that passes through a specified point. That is to say, we shall be concerned with numerical methods that will determine an approximate solution to the IVP

$$\left. \begin{aligned} x'(t) &= f(t, x(t)), \qquad t > t_0 \\ x(t_0) &= \eta \end{aligned} \right\}. \tag{1.6}$$

For example, the heavier curves emanating from the points P shown in Figure 1.1 are the solutions of the IVPsfor the differential equations (1.1) and (1.2) in Example 1.1 with the common starting value $x(\frac{5}{2}) = 5$. We emphasize that approximate solutions will only be computed to IVPs—general solutions containing arbitrary constants cannot be approximated by the techniques described in this book. Because most ODEs describe the temporal rate of change of some physical quantity, we will refer to t as the *time* variable.

It will always be assumed that the IVP has a solution in some time interval $[t_0, t_f]$ for some value of $t_f > t_0$. In many cases there is a solution for all $t > t_0$ but we are usually interested in the solution only for a limited time, perhaps up until it approaches a steady state—after which time the solution changes imperceptibly. In some instances (such as those modelling explosions, for instance) the solution may not exist for all time. See, for instance, Exercise 1.3. The reference texts listed in the Preface provide more information regarding existence and uniqueness theory for ODEs.

Problems of the type (1.6) involving just a single first-order differential equation form only a small class of the ODEs that might need to be solved in practical applications. However, the methods we describe will also be applicable to systems of differential equations and thereby cover many more possibilities.

1.1 Systems of ODEs

Consider the IVP for the pair of ODEs

$$u'(t) = p(t, u, v), \qquad u(t_0) = \eta_0,$$
$$v'(t) = q(t, u, v), \qquad v(t_0) = \eta_1, \qquad (1.7)$$

where $u = u(t)$ and $v = v(t)$. Generally these are coupled equations that have to be solved simultaneously. They can be written as a vector system of differential equations by defining

$$\boldsymbol{x} = \begin{bmatrix} u \\ v \end{bmatrix}, \quad \boldsymbol{f}(t, \boldsymbol{x}) = \begin{bmatrix} p(t, u, v) \\ q(t, u, v) \end{bmatrix}, \quad \boldsymbol{\eta} = \begin{bmatrix} \eta_0 \\ \eta_1 \end{bmatrix},$$

so that we have the vector form of (1.6):

$$\left. \begin{array}{l} \boldsymbol{x}'(t) = \boldsymbol{f}(t, \boldsymbol{x}(t)), \qquad t > t_0 \\ \boldsymbol{x}(t_0) = \boldsymbol{\eta} \end{array} \right\}. \qquad (1.8)$$

In this case, $\boldsymbol{x}, \boldsymbol{\eta} \in \mathbb{R}^2$ and $\boldsymbol{f} : \mathbb{R} \times \mathbb{R}^2 \to \mathbb{R}^2$. The convention in this book is that bold-faced symbols, such as $\boldsymbol{x}$ and $\boldsymbol{f}$, are used to distinguish vector-valued quantities from their scalar counterparts shown in normal font: x, f.

More generally, IVPsfor coupled systems of m first-order ODEs will also be written in the form (1.8) and in such cases $\boldsymbol{x}$ and $\boldsymbol{\eta}$ will represent vectors in $\mathbb{R}^m$ and $\boldsymbol{f} : \mathbb{R} \times \mathbb{R}^m \to \mathbb{R}^m$.

Differential equations in which the time variable does not appear explicitly are said to be *autonomous*. Thus, $x'(t) = g(x(t))$ and $x'(t) = x(t)(1 - x(t))$ are autonomous but $x'(t) = (1 - 2t)x(t)$ is not. Non-autonomous ODEs can be rewritten as autonomous systems—we shall illustrate the process for the two-variable system (1.7). We now define

$$\boldsymbol{x} = \begin{bmatrix} t \\ u \\ v \end{bmatrix}, \quad \boldsymbol{f}(\boldsymbol{x}) = \begin{bmatrix} 1 \\ p(t, u, v) \\ q(t, u, v) \end{bmatrix}, \quad \boldsymbol{\eta} = \begin{bmatrix} t_0 \\ \eta_0 \\ \eta_1 \end{bmatrix}.$$

The first component x_1 of $\boldsymbol{x}$ satisfies $x_1'(t) = 1$ with $x_1(0) = t_0$ and so $x_1(t) \equiv t$. We now have the autonomous IVP

$$\left. \begin{array}{l} \boldsymbol{x}'(t) = \boldsymbol{f}(\boldsymbol{x}(t)), \qquad t > t_0 \\ \boldsymbol{x}(t_0) = \boldsymbol{\eta} \end{array} \right\}. \qquad (1.9)$$

Example 1.3 (Lotka–Volterra Equations)

The Lotka–Volterra equations (developed circa 1925) offer a simplistic model of
the conflict between populations of predators and prey. Suppose, for instance,
that the numbers of rabbits and foxes in a certain region at time t are $u(t)$ and
$v(t)$ respectively. Then, given the initial population sizes $u(0)$ and $v(0)$ at time
$t = 0$, their numbers might evolve according to the autonomous system

$$
\begin{aligned}
u'(t) &= 0.05u(t)\big(1 - 0.01v(t)\big), \\
v'(t) &= 0.1v(t)\big(0.005u(t) - 2\big).
\end{aligned}
\tag{1.10}
$$

These ODEs reproduce certain features that make sense in this predator-prey
situation. In the first of these:

- Increasing the number of foxes $(v(t))$ decreases the rate $(u'(t))$ at which
  rabbits are produced (because more rabbits get eaten).

- Increasing the number of rabbits $(u(t))$ increases the rate at which rabbits
  are produced (because more pairs of rabbits are available to mate).

The solutions corresponding to an initial population of 1500 rabbits (solid
curves) and 100 foxes (dashed curves) are shown on the left of Figure 1.3.
On the right we plot these, and other solutions, in the phase plane, i.e., the
locus of the point $(u(t), v(t))$ parameterized by t for $0 \le t \le 600$. Initially there
are 100 foxes and the three curves correspond to there being initially 600, 1000,
and 1500 rabbits (the starting values are shown by solid dots). The periodic
nature of the populations can be deduced from the fact that these are closed
curves. The "centre" of rotation is indicated by a small circle.   □

The text by Murray [56] is packed with examples where ODEs are used to
model biological processes.

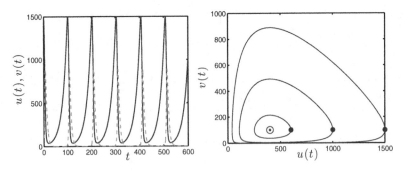

Fig. 1.3 Solutions for the Lotka–Volterra equations (1.10)

Example 1.4 (Biochemical Reactions)

In biochemistry, a *Michaelis–Menten*-type process involving

– a substrate S,

– an enzyme E,

– a complex C, and

– a product P

could be summarized through the reactions

$$S + E \xrightarrow{c_1} C$$
$$C \xrightarrow{c_2} S + E$$
$$C \xrightarrow{c_3} P + E.$$

In the framework of chemical kinetics, this set of reactions may be interpreted as the ODE system

$$\begin{aligned}
S'(t) &= -c_1 S(t)\, E(t) + c_2 C(t), \\
E'(t) &= -c_1 S(t)\, E(t) + (c_2 + c_3) C(t), \\
C'(t) &= c_1 S(t)\, E(t) - (c_2 + c_3) C(t), \\
P'(t) &= c_3 C(t),
\end{aligned}$$

where $S(t)$, $E(t)$, $C(t)$ and $P(t)$ denote the concentrations of substrate, enzyme, complex and product, respectively, at time t.

Letting $\boldsymbol{x}(t)$ be the vector $[S(t), E(t), C(t), P(t)]^{\mathrm{T}}$, this system fits into the general form (1.8) with $m = 4$ components and

$$\boldsymbol{f}(t, \boldsymbol{x}) = \begin{bmatrix}
-c_1 x_1 x_2 + c_2 x_3 \\
-c_1 x_1 x_2 + (c_2 + c_3) x_3 \\
c_1 x_1 x_2 - (c_2 + c_3) x_3 \\
c_3 x_3
\end{bmatrix}.$$

In Figure 1.4 we show how the levels of substrate and product, $x_1(t)$ and $x_4(t)$ respectively, evolve over time. It is clear that the substrate becomes depleted as product is created. Here we took initial conditions of $x_1(0) = 5 \times 10^{-7}$, $x_2(0) = 2 \times 10^{-7}$, $x_3(0) = x_4(0) = 0$, with rate constants $c_1 = 10^6$, $c_2 = 10^{-4}$, $c_3 = 0.1$, based on those in Wilkinson [69] that were also used in Higham [33].

We refer to Alon [1] and Higham [33] for more details about how ODE models are used in chemistry and biochemistry. □

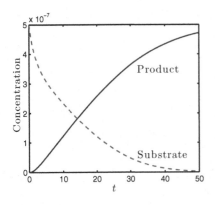

Fig. 1.4 Substrate and product concentrations from the chemical kinetics ODE of Example 1.4

Example 1.5 (Fox-Rabbit Pursuit)

Curves of pursuit arise naturally in military and predator-prey scenarios whenever there is a moving target. Imagine that a rabbit follows a predefined path, $(r(t), s(t))$, in the plane in an attempt to shake off the attentions of a fox. Suppose further that the fox runs at a speed that is a constant factor k times the speed of the rabbit, and that the fox chases in such a way that at all times its tangent points at the rabbit. Straightforward arguments then show that the fox's path $(x(t), y(t))$ satisfies

$$x'(t) = R(t) \ (r(t) - x(t)),$$
$$y'(t) = R(t) \ (s(t) - y(t)),$$

where

$$R(t) = \frac{k\sqrt{r'(t)^2 + s'(t)^2}}{\sqrt{(r(t) - x(t))^2 + (s(t) - y(t))^2}}.$$

In the case where the rabbit's path $(r(t), s(t))$ is known, this is an ODE system of the form (1.8) with $m = 2$ components.

In Figure 1.5 the rabbit follows an outward spiral

$$\begin{bmatrix} r(t) \\ s(t) \end{bmatrix} = \sqrt{1+t} \begin{bmatrix} \cos t \\ \sin t \end{bmatrix},$$

shown as a dashed curve in the x, y plane, with a solid dot marking the initial location. The fox's path, found by solving the ODE with a numerical method, is shown as a solid line. The fox is initially located at $x(0) = 3$, $y(0) = 0$ and travels $k = 1.1$ times as quickly as the rabbit. We solved the ODE up to $t = 5.0710$, which is just before the critical time where rabbit and fox meet (marked by an asterisk). At that point the ODE would be ill defined because of a division-by-zero error in the definition of $R(t)$ (and because of a lack of rabbit!).

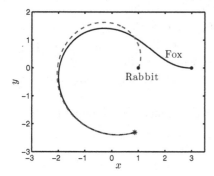

Fig. 1.5 Fox-rabbit pursuit curve from Example 1.5

This example is taken from Higham and Higham [34, Chapter 12], where further details are available. A comprehensive and very entertaining coverage of historical developments in the field of pursuit curves can be found in Nahin's book [58].   □

Example 1.6 (Zombie Outbreak)

ODEs are often used in epidemiology and population dynamics to describe the spread of a disease. To illustrate these ideas in an eye-catching science fiction context, Munz et al. [55] imagined a zombie outbreak. At each time t, their model records the levels of

– humans $H(t)$,

– zombies $Z(t)$,

– removed ('dead' zombies) $R(t)$, which may return as zombies.

It is assumed that a zombie may irreversibly convert a human into a zombie. On the other hand, zombies cannot be killed, but a plucky human may temporarily send a zombie into the 'removed' class. The simplest version of the model takes the form

$$
\begin{aligned}
H'(t) &= -\beta H(t)Z(t), \\
Z'(t) &= \beta H(t)Z(t) + \zeta R(t) - \alpha H(t)Z(t), \\
R'(t) &= \alpha H(t)Z(t) - \zeta R(t).
\end{aligned}
$$

The (constant, positive) parameters in the model are:

– α, dealing with human–zombie encounters that remove zombies.

– β, dealing with human–zombie encounters that convert humans to zombies.

– ζ, dealing with removed zombies that revert to zombie status.

Figure 1.6 shows the case where $\beta = 0.01$. $\alpha = 0.005$ and $\zeta = 0.02$. We set the initial human population to be $H(0) = 500$ with $Z(0) = 10$ zombies. We have in mind the scenario where a group of zombies from Lenzie, near Glasgow, attacks the small community of Newport in Fife. The figure shows the evolution of the human and zombie levels, and the doomsday outcome of total zombification.

The issue of how to infer the model parameters from observed human/zombie population levels is treated in [7].

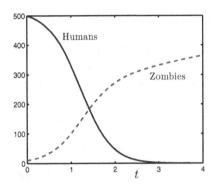

Fig. 1.6 Human and zombie population levels from Example 1.6

Example 1.7 (Method of Lines for Partial Differential Equations)

The numerical solution of partial differential equations (PDEs) can lead to extremely large systems of ODEs. Consider the solution of Fisher's equation

$$\frac{\partial u}{\partial t} = \frac{\partial^2 u}{\partial x^2} + u(1 - u)$$

defined for $0 < x < 5$ and $t > 0$. The solution $u(t, x)$ also has to satisfy boundary conditions $u(t, 0) = 0$ and $u(t, 5) = 1$, and match given initial data $u(0, x)$, which we take to be

$$u(0, x) = e^{x/5 - 1} \sin^2 \tfrac{3}{10} \pi x, \qquad 0 \le x \le 5.$$

This is a simple example of one of the many initial-boundary-value problems for equations of *reaction-diffusion* type that occur in chemistry, biology, ecology, and countless other areas. It can be solved numerically by dividing the interval $0 < x < 5$ into $5N$ subintervals (say) by the points $x_j = j/N$, $j = 0 : 5N$ and using $u_j(t)$ to denote the approximation to $u(x_j, t)$ at each of these locations.[3] After approximating the second derivative in the equation by finite differences

---

[3]For integers $m < n$, $j = m : n$ is shorthand for $j = m, m + 1, m + 2, \ldots, n$.

(see, for instance, Leveque [47] or Morton and Mayers [54]) we arrive at a system of $5N - 1$ differential equations, of which a typical equation is

$$u'_j = N^2(u_{j-1} - 2u_j + u_{j+1}) + u_j - u_j^2,$$

$(j = 1 : 5N-1)$ with end conditions $u_0(t) = 0$, $u_{5N}(t) = 1$ and initial conditions that specify the values of $u_j(0)$:

$$u_j(0) = e^{1-j/5N} \sin^2 \tfrac{3j}{10N}\pi, \qquad j = 1 : 5N - 1.$$

By defining the vector function of t

$$\boldsymbol{u}(t) = [u_1(t), u_2(t), \dots, u_{5N-1}(t)]^{\mathrm{T}}$$

and the $(5N - 1) \times (5N - 1)$ tridiagonal matrix A

$$A = N^2 \begin{bmatrix} -2 & 1 & & & \\ 1 & -2 & 1 & & \\ & \ddots & \ddots & \ddots & \\ & & 1 & -2 & 1 \\ & & & 1 & -2 \end{bmatrix} \tag{1.11}$$

we obtain a system of the form

$$\boldsymbol{u}'(t) = A\boldsymbol{u}(t) + \boldsymbol{g}(\boldsymbol{u}(t)), \qquad t > 0,$$

where the jth component of $\boldsymbol{g}(\boldsymbol{u})$ is $u_j - u_j^2$ (except for the last component, $g_{5N-1}(\boldsymbol{u})$, which has an additional term, N^2, due to the non-zero boundary condition at $x = 5$) and $\boldsymbol{u}(0) = \boldsymbol{\eta}$ is the known vector of initial data. The solution with $N = 6$ is shown in Figure 1.7 for $0 \le t \le 0.1$. □

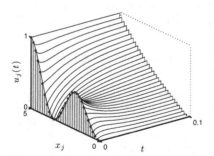

Fig. 1.7 Solution for the 29 ODEs approximating Fisher's equation with $N = 6$. The initial conditions for the components are indicated by solid dots ($\bullet$)

Generally, PDEs involving one spatial variable (x) typically lead to 100–1000 ODEs. With two spatial variables (x and y) these numbers become 100^2–1000^2 so the number of ODEs in a system may run to millions.

1.2 Higher Order Differential Equations

Some mathematical models involve derivatives higher than the first. For example, Newton's laws of motion involve *acceleration*—the second derivative of position with respect to time. Such higher-order ODEs are automatically contained in the framework of *systems of first-order* ODEs. We first illustrate the idea with an example.

Example 1.8 (Van der Pol Oscillator)

For the second-order IVP

$$x''(t) + 10(1 - x^2(t))x'(t) + x(t) = \sin \pi t,$$
$$x(t_0) = \eta_0, \qquad x'(t_0) = \eta_1,$$

we define $u = x$, $v = x'$ so that

$$u'(t) = v(t),$$
$$v'(t) = -10(1 - u^2(t))v(t) - u(t) + \sin \pi t,$$

with initial conditions $u(t_0) = \eta_0$ and $v(t_0) = \eta_1$. These equations are now in the form (1.7), so that we may again write them as (1.8), where

$$\boldsymbol{x}(t) = \begin{bmatrix} u(t) \\ v(t) \end{bmatrix}, \quad \boldsymbol{f}(t, \boldsymbol{x}(t)) = \begin{bmatrix} v(t) \\ -10(1 - u^2(t))v(t) - u(t) + \sin \pi t \end{bmatrix}.$$

The solution of the unforced equation, where the term $\sin \pi t$ is dropped, for the initial condition $u(0) = 1$, $v(0) = 5$ is shown in Figure 1.8 for $0 \le t \le 100$. After a short initial phase ($0 \le t \le 0.33$—shown as a broken curve emanating from P) the solution approaches a periodic motion; the limiting closed curve is known as a *limit cycle*. In the figure on the left the solution $v(t) \equiv x'(t)$ is shown as a broken curve and appears as a sequence of spikes. $\qquad\square$

It is straightforward to extend the second-order example in Example 1.8 to a differential equation of order m:

$$x^{(m)}(t) = f\big(t, x(t), x'(t), \ldots, x^{(m-1)}(t)\big). \tag{1.12}$$

New dependent variables may be introduced for the function $x(t)$ and each of its first $(m - 1)$ derivatives:

$$x_1(t) = x(t),$$
$$x_2(t) = x'(t),$$
$$\vdots$$
$$x_m(t) = x^{(m-1)}(t).$$

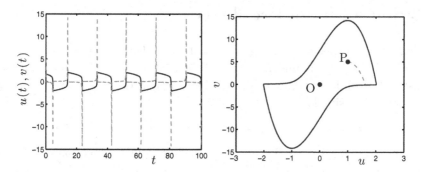

Fig. 1.8 Solution for the ODEs representing the unforced Van der Pol equation of Example 1.8. On the left, $u(t) \equiv x(t)$ (solid curve) and $v(t) \equiv x'(t)$ (broken curve) are shown as functions of t. The phase portrait is drawn on the right, where P marks the initial point

There are obvious relationships between the first $(m-1)$ such functions:

$$x_1'(t) = x_2(t),$$
$$x_2'(t) = x_3(t),$$
$$\vdots$$
$$x_{m-1}'(t) = x_m(t).$$

These, together with the differential equation, which has become,

$$x_m'(t) = f(t, x_1(t), x_2(t), \ldots, x_m(t)),$$

give a total of m ODEs for the components of the m-dimensional vector function $\boldsymbol{x} = [x_1, x_2, \ldots, x_m]^{\mathrm{T}}$. The corresponding IVP for $x(t)$ has values specified for the function itself as well as its first $(m-1)$ derivatives at the initial time $t = t_0$, and these give directly the initial conditions for the m components of $\boldsymbol{x}(t_0)$. We therefore have an IVP of the form (1.8), where

$$\boldsymbol{f}(t, \boldsymbol{x}(t)) = \begin{bmatrix} x_2(t) \\ x_3(t) \\ \vdots \\ x_m(t) \\ f(t, x_1(t), x_2(t), \ldots, x_m(t)) \end{bmatrix}, \qquad \boldsymbol{\eta} = \begin{bmatrix} x(0) \\ x'(0) \\ \vdots \\ x^{(m-2)}(0) \\ x^{(m-1)}(0) \end{bmatrix}.$$

The general principle in this book is that numerical methods will be constructed and analysed, in the first instance, for approximating the solutions of IVPs for scalar problems of the form given in (1.6), after which we will discuss how the same methods apply to problems in the vector form (1.8).

1.3 Some Model Problems

One of our strategies in assessing numerical methods is to apply them to very simple problems to gain understanding of their behaviour and then attempt to translate this understanding to more realistic scenarios.

Examples of simple model ODEs are:

1. $x'(t) = \lambda x(t)$, $x(0) = 1$, $\lambda \in \Re$; usually $\lambda < 0$ so that $x(t) \to 0$ as $t \to \infty$.

2. $x'(t) = \mathrm{i}x(t)$, $x(0) = 1$ ($\mathrm{i} = \sqrt{-1}$) modelling oscillatory motion (see Exercise 1.6).

3. $x'(t) = -100x(t) + 100\mathrm{e}^{-t}$, $x(0) = 2$. The exact solution is $x(t) = \mathrm{e}^{-t} + \mathrm{e}^{-100t}$, which combines two decaying terms, one of which is very rapid and one that decays more slowly. Hence, there are two distinct time scales in this solution.[4]

4. $x'(t) = 1$, $x(0) = 0$.

5. $x'(t) = 0$, $x(0) = 0$.

The last two, in particular, are trivial, but of course, a numerical method, if it is to be useful, must work well on such simple examples—if methods cannot reproduce the solutions to these problems then they are deemed to be unsuitable for solving *any* IVPs. Through use of simple ODEs we can highlight deficiencies; and the simpler the ODE, the simpler the analysis is.

Example 1.9 (A Cooling Cup of Coffee)

Although linear ODEs such as 1–5 above are useful for testing numerical methods, they may also arise as mathematical models. For example, suppose that a cup of boiling coffee is prepared at time $t = 0$ and cools according to Newton's law of cooling: the rate of change of temperature is proportional to the difference in temperature between the coffee and the surrounding room (see, for instance, Chapter 12 of Fulford et al. [20]). Suppose that $u(t)$ represents the coffee temperature (in degrees Celsius) after t hours. This leads to the differential equation

$$u'(t) = -\alpha(u(t) - v),$$

where α is known as the rate constant (which will be taken to be $\alpha = 8°\mathrm{C}\,\mathrm{h}^{-1}$) and v represents room temperature. We consider a number of scenarios (see Figure 1.9).

---

[4]Recall that the function $A\mathrm{e}^{-\lambda t}$ ($\lambda \in \mathbb{R}$) decays to half its initial value (A at $t = 0$) in a time $t = (\log 2)/\lambda \approx 0.7/\lambda$, commonly referred to as its half-life. The larger the rate constant λ is, the more quickly it decays.

1. Room temperature is a constant $v = 20°C$. We then have a scalar IVP

$$u'(t) = -8(u(t) - 20), \qquad u(0) = 100. \tag{1.13}$$

   This has solution $u(t) = 80e^{-8t} + 20$.

2. The room is also cooling according to Newton's law from an initial temperature of $20°C$ with a rate constant $1/8$ and exterior temperature of $5°C$. With $v(t)$ denoting room temperature, we have the following system of two ODEs

$$\begin{aligned} u'(t) &= -8(u(t) - v(t)), & u(0) &= 100, \\ v'(t) &= -(v(t) - 5)/8, & v(0) &= 20. \end{aligned} \tag{1.14}$$

   The second of these ODEs may be solved to give

$$v(t) = 15e^{-t/8} + 5,$$

   which can be used to give a scalar IVP for u:

$$u'(t) = -8(u(t) - 15e^{-t/8} - 5), \qquad u(0) = 100. \tag{1.15}$$

   It follows that $u(t) = \frac{1675}{21}e^{-8t} + \frac{320}{21}e^{-t/8} + 5$ (see Exercise 1.13).

3. There is a third situation where the coffee container is well insulated and the room is not. We may model this by the IVP

$$u'(t) = -\tfrac{1}{8}(u(t) - 5 + 5025e^{-8t}), \qquad u(0) = 100, \tag{1.16}$$

so that it has the same solution as (1.15). Although the two problems have the same solution, numerical methods applied to the two problems will generally behave differently, and this will be explored in due course (see Examples 6.1 and 6.2). □

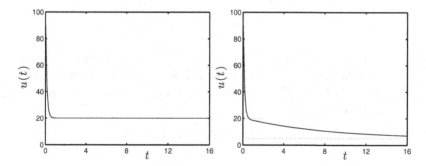

Fig. 1.9   The solution of (1.13) (left) and (1.14)–(1.16) (right)

EXERCISES

1.1.* Use the variation of constants formula (1.4) to derive the exact solution of the ODE (1.1).

1.2.** Complete the details leading up to the variation of constants formula (1.4).

1.3.** Show that the IVP $x'(t) = x(t)(1 - x(t))$, $x(10) = -1/5$, has solution $x(t) = 1/(1 - 6e^{10-t})$. Deduce that $x(t) \to -\infty$ as $t \to 10 + \log 6 \approx 11.79$—the solution is said to blow up in finite time.

1.4.** Show that $x(t) = t(4 - t)/4$ satisfies both the ODE $x'(t) = \sqrt{1 - x(t)}$ and the starting condition $x(0) = 0$ but cannot be a solution of the IVP for $t > 2$ since the formula for $x(t)$ has a negative derivative while the right-hand side of the ODE is non-negative for all t.

1.5.** Rewrite the following IVPsas IVPs for first order systems, as illustrated in Section 1.2.

(a) Simple harmonic motion:

$$\theta''(t) + \theta(t) = 0, \qquad \theta(0) = \pi/10, \quad \theta'(0) = 0.$$

(b)

$$x''(t) - x'(t) - 2x(t) = 1 + 2t, \qquad x(0) = 1, \quad x'(0) = 1.$$

(c)

$$x'''(t) - 2x''(t) - x'(t) + 2x(t) = 1 - 2t,$$
$$x(0) = 1, \quad x'(0) = 1, \quad x''(0) = 0.$$

1.6.** Suppose that $\lambda \in \mathbb{C}$. Show that the complex ODE $x'(t) = \lambda x(t)$ may be written as a first-order system of two real ODEs. (Hint: let $x(t) = u(t) + iv(t)$ and $\lambda = a + ib$ and consider the real and imaginary parts of the result.)

When $a = 0$, use the chain rule to verify that

$$\frac{\mathrm{d}}{\mathrm{d}t}\left(u^2(t) + v^2(t)\right) = 0$$

so that solutions lie on the family of circles $u^2(t) + v^2(t) = \text{constant}$ in the u-v phase plane.

1.7.* Write the differential equation of Example 1.2 as an autonomous system with two components.

1.8.* Use the system of ODEs (1.10) to decide whether the periodic motion described by the rightmost set of curves in Figure 1.3 is clockwise or counter clockwise.

1.9.** Show that the IVP

$$x''(t) - ax'(t) - bx(t) = f(t), \quad x(0) = \xi, \quad x'(0) = \eta$$

may be written as a first-order system $x'(t) = Ax(t) + g(t)$, where A is a 2×2 matrix, $x = [x, x']^T$ and the components of the 2×1 vector g should be related to the forcing function $f(t)$. What is the characteristic polynomial of A?

1.10.** Write the IVP

$$x'''(t) - ax''(t) - bx'(t) - cx = f(t),$$
$$x(0) = \xi, \quad x'(0) = \eta, \quad x''(0) = \zeta$$

as a first-order system $x'(t) = Ax(t) + g(t)$. What is the characteristic polynomial of A?

1.11.* By summing the first, third and fourth ODEs in the Michaelis–Menten chemical kinetics system of Example 1.4, show that $S'(t) + C'(t) + P'(t) = 0$. This implies that the total amount of "substrate plus complex plus product" does not change over time. Convince yourself that this *conservation law* is intuitively reasonable, given the three chemical reactions involved. (Invariants of this type will be discussed in Chapter 14.)

1.12.* Show that, for the populations in Example 1.6, the total population $H(t) + Z(t) + R(t)$ remains constant in time.

1.13.* Differentiate $u(t) = \frac{1675}{21}e^{-8t} + \frac{320}{21}e^{-t/8} + 5$ and hence verify that this function solves both the ODEs (1.15) and (1.16).

1.14.*** Suppose that $x(t)$ denotes the solution of the ODE $x'(t) = 1 + x^2(t)$.

(a) Find the general solution of the given differential equation and show that it contains one arbitrary constant.

(b) Use the change of dependent variable $x(t) = -y'(t)/y(t)$ to show that $x(t)$ will solve the given first-order differential equation provided $y(t)$ solves a certain linear second-order differential equation and determine the general solution for $y(t)$.

Deduce the general solution for $x(t)$ and explain why this appears to contain two arbitrary constants.

1.15.* Consider the logistic equation $x'(t) = ax(t)\left(1 - \frac{x(t)}{X}\right)$, in which the positive constants a and X are known as the growth (proliferation) rate and carrying capacity, respectively. This might model the number of prey $x(t)$ at time t in a total population of predators and prey comprising a fixed number X of individuals.

Derive the corresponding ODE for the fraction $u(\tau) = x(t)/X$ of prey in the population as a function of the scaled time $\tau = a\,t$.

2

Euler's Method

During the course of this book we will describe three families of methods for numerically solving IVPs: the Taylor series (TS) method, linear multistep methods (LMMs) and Runge–Kutta (RK) methods.

They aim to solve *all* IVPs of the form

$$\left. \begin{array}{l} x'(t) = f(t, x(t)), \quad t > t_0 \\ x(t_0) = \eta \end{array} \right\} \tag{2.1}$$

that possess a unique solution on some specified interval, $t \in [t_0, t_f]$ say.

The three families can all be interpreted as generalisations of the world's simplest method: Euler's method. It is appropriate, therefore, that we start with detailed descriptions of Euler's method and its derivation, how it can be applied to systems of ODEs as well as to scalar problems, and how it behaves numerically.

We shall, throughout, use h to refer to a "small" positive number called the "step size" or "grid size": we will seek approximations to the solution of the IVP at particular times $t = t_0, t_0 + h, t_0 + 2h, ..., t_0 + nh, ...$, i.e., approximations to the sequence of numbers $x(t_0), x(t_0 + h), x(t_0 + 2h), ..., x(t_0 + nh), ...$, rather than an approximation to the curve $\{x(t) : t_0 \leq t \leq t_f\}$.

D.F. Griffiths, D.J. Higham, *Numerical Methods for Ordinary Differential Equations*,
Springer Undergraduate Mathematics Series, DOI 10.1007/978-0-85729-148-6_2,
© Springer-Verlag London Limited 2010

2.1 A Preliminary Example

To illustrate the process we begin with a specific example before treating the general case.

Example 2.1

Use a step size $h = 0.3$ to develop an approximate solution to the IVP

$$\left. \begin{array}{l} x'(t) = (1 - 2t)x(t), \quad t > 0 \\ \quad\quad x(0) = 1 \end{array} \right\} \tag{2.2}$$

over the interval $0 \le t \le 0.9$.

We have purposely chosen a problem with a known exact solution

$$x(t) = \exp[\tfrac{1}{4} - (\tfrac{1}{2} - t)^2] \tag{2.3}$$

so that we can more easily judge the level of accuracy of our approximations.

At $t = 0$ we have $x(0) = 1$ and, from the ODE, $x'(0) = 1$. This information enables us to construct the tangent line to the solution curve at $t = 0$: $x = 1 + t$. At the specified value $t = 0.3$ we take the value $x = 1.3$ as our approximation: $x(0.3) \approx 1.3$. This is illustrated on the left of Figure 2.1—here the line joining $P_0(0,1)$ and $P_1(0.3, 1.3)$ is tangent to the exact solution at $t = 0$.

To organize the results we let $t_n = nh$, $n = 0, 1, 2, \ldots$, denote the times at which we obtain approximate values and denote by x_n the computed approximation to the exact solution $x(t_n)$ at $t = t_n$. It is also convenient to record the value of the right side of the ODE at the point (t_n, x_n):

$$x_n' = (1 - 2t_n)x_n.$$

The initial conditions and first step are summarized by

$$n = 0: \quad t_0 = 0, \qquad n = 1: \quad t_1 = t_0 + h = 0.3,$$
$$x_0 = 1, \qquad\qquad\qquad x_1 = x_0 + hx_0' = 1.3,$$
$$x_0' = 1, \qquad\qquad\qquad x_1' = (1 - 2t_1)x_1 = 0.52.$$

The process is now repeated: we construct a line through $P_1(t_1, x_1)$ that is tangent to the solution of the differential equation that passes through P_1 (shown as a dashed curve in Figure 2.1 (m iddle)). This tangent line passes through (t_1, x_1) and has slope x_1':

$$x = x_1 + (t - t_1)x_1'.$$

At $t = 2h$ we find

$$x_2 = x_1 + hx_1',$$

so the calculations for the next two steps are

$n = 2:$ $t_2 = t_1 + 0.3 = 0.6,$ $n = 3:$ $t_3 = t_2 + h = 0.9,$

           $x_2 = x_1 + hx_1' = 1.456,$ $x_3 = x_2 + hx_2' = 1.3686,$

           $x_2' = (1 - 2t_2)x_2 = -0.2912,$ $x_3' = (1 - 2t_3)x_3 = -1.0949.$

The points $P_n(t_n, x_n)$ $(n = 0, 1, 2, 3)$ are shown in Figure 2.1. For each n the line P_nP_{n+1} is tangent to the solution of the differential equation $x'(t) = (1-2t)x(t)$ that passes through the point $x(t) = x_n$ at $t = t_n$. $\Box$

2.1.1 Analysing the Numbers

The computed points P_1, P_2, P_3 in Figure 2.1 are an appreciable distance from the solid curve representing the exact solution of our IVP. This is a consequence of the fact that our chosen time step $h = 0.3$ is too large. In Figure 2.2 we show the results of computing the numerical solutions (shown as solid dots) over the extended interval $0 \le t \le 3$ with $h = 0.3, 0.15$, and 0.075. Each time h is halved:

1. Twice as many steps are needed to find the solution at $t = 3$.

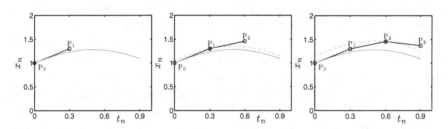

Fig. 2.1 The development of a numerical solution to the IVP in Example 2.1 over three time steps. The exact solution of the IVP is shown as a solid curve

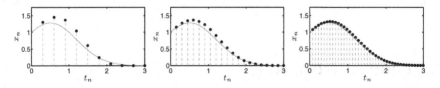

Fig. 2.2 Numerical solutions for Example 2.1 with $h = 0.3$ (left), $h = 0.15$ (middle) and $h = 0.075$ (right)

h	x_n	Global errors (GEs)	GE/h
0.3	$x_3 = 1.3686$	$x(0.9) - x_3 = -0.2745$	-0.91
0.15	$x_6 = 1.2267$	$x(0.9) - x_6 = -0.1325$	-0.89
0.075	$x_{12} = 1.1591$	$x(0.9) - x_{12} = -0.0649$	-0.86
Exact	$x(0.9) = 1.0942$		

Table 2.1 Numerical results at $t = 0.9$ with $h = 0.3, 0.15$ and 0.075

2. The computed points lie closer to the exact solution curve. This illustrates the notion that the numerical solution *converges* to the exact solution as $h \to 0$.

To obtain more concrete evidence, the numerical solutions at $t = 0.9$ are shown in Table 2.1; since the exact solution at this time is known, the errors in the approximate values may be calculated. The difference

$$e_n = x(t_n) - x_n$$

is referred to as the *global error* (GE) at $t = t_n$. It is seen from the table that halving h results in the error being approximately halved. This suggests that the GE is proportional to h:

$$e_n \propto h.$$

It is possible to prove that this is true for a wide class of (nonlinear) IVPs; we will develop some theory along these lines in Section 2.4 for a model linear problem. The final column in Table 2.1 suggests that the constant of proportionality in this case is about $-0.9 : e_n \approx -0.9h$, when $nh = 0.9$; so, were we to require an accuracy of three decimal places, h would have to be small enough that $|e_n| < 0.0005$, i.e. $h < 0.0005/0.9 \approx 0.00055$. Consequently, the integration to $t = 0.9$ would require about $n = 0.9/h \approx 1620$ steps.

An alternative view is that, for each additional digit of accuracy, the GE should be reduced by a factor of 10, so h should also be reduced by a factor of 10 and, consequently, 10 times as many steps are required to integrate to the same final time. Thus, 10 times as much computational effort has to be expended to improve the approximation by just one decimal place—a substantial increase in cost.

2.2 Landau Notation

We will make extensive use of the standard notation wherein $\mathcal{O}(h^p)$, with p a positive integer, refers to a quantity that decays at least as quickly as h^p when

h is small enough. More precisely, we write $z = \mathcal{O}(h^p)$ if there exist positive constants h_0 and C such that

$$|z| \leq Ch^p, \quad \text{for all } 0 < h < h_0,$$

so that z converges to zero as $h \to 0$ and the order (or rate) of convergence is p. The $\mathcal{O}(h^p)$ notation is intended to convey the impression that we are interested primarily in p and not C, and we say that "z is of order h^p" or "z is of pth order". For example, the Maclaurin expansion of e^h is

$$e^h = 1 + h + \tfrac{1}{2!}h^2 + \tfrac{1}{3!}h^3 + \cdots + \tfrac{1}{n!}h^n + \cdots ,$$

from which we deduce

$$e^h = 1 + \mathcal{O}(h) \tag{2.4a}$$

$$= 1 + h + \mathcal{O}(h^2) \tag{2.4b}$$

$$= 1 + h + \tfrac{1}{2!}h^2 + \mathcal{O}(h^3). \tag{2.4c}$$

We use this idea to compare the magnitudes of different terms: for example, the $\mathcal{O}(h^3)$ term in (2.4c) will always be *smaller* than the $\mathcal{O}(h^2)$ term in (2.4b) provided h is sufficiently small; the higher the order, the smaller the term. The notion of "sufficiently small" is not precise in general: h^3 is less than h^2 for $h < 1$, while $100h^3$ is less than h^2 for $h < 1/100$. Thus, the smallness of h depends on the context.

All methods that we discuss will be founded on the assumption that the solution $x(t)$ is smooth in the sense that as many derivatives as we require are continuous on the interval (t_0, t_f). This will allow us to use as many terms as we wish in Taylor series expansions.

2.3 The General Case

To develop Euler's method for solving the general IVP (2.1) the approximation process begins by considering the Taylor series of $x(t + h)$ with remainder (see Appendix B):

$$x(t + h) = x(t) + hx'(t) + R_1(t). \tag{2.5}$$

The remainder term $R_1(t)$ is called the *local truncation error* (LTE). If $x(t)$ is twice continuously differentiable on the interval (t_0, t_f) the remainder term may be written as

$$R_1(t) = \tfrac{1}{2!}h^2 x''(\xi), \quad \xi \in (t, t + h). \tag{2.6}$$

Then, if a positive number M exists so that $|x''(t)| \leq M$ for all $t \in (t_0, t_f)$, it follows that

$$|R_1(t)| \leq \tfrac{1}{2} M h^2,$$

i.e., $R_1(t) = \mathcal{O}(h^2)$.

To derive Euler's method we begin by substituting $x'(t) = f(t, x)$ into the Taylor series (2.5) to obtain

$$x(t + h) = x(t) + h f(t, x(t)) + R_1(t). \tag{2.7}$$

We now introduce a grid of points $t = t_n$, where

$$t_n = t_0 + nh, \qquad n = 1 : N \tag{2.8}$$

and $N = \lfloor (t_f - t_0)/h \rfloor$ is the number of steps[1] of length h needed to reach, but not exceed, $t = t_f$. With $t = t_n$ (for $n < N$) in (2.7) we have

$$x(t_{n+1}) = x(t_n) + h f(t_n, x(t_n)) + R_1(t_n), \qquad n = 0 : N - 1,$$

and, with the initial condition $x(t_0) = \eta$, we would be able to compute the exact solution of the IVP on the grid $\{t_n\}_{n=0}^{N}$ using this recurrence relation were the LTE term $R_1(t)$ not present.

However, since $R_1(t) = \mathcal{O}(h^2)$, it can be made arbitrarily small (by taking h to be sufficiently small) and, when neglected, we obtain *Euler's* method,

$$x_{n+1} = x_n + h f(t_n, x_n), \qquad n = 0, 1, 2, \ldots,$$

with which we can compute the sequence $\{x_n\}$ given that $x_0 = \eta$. We shall use the notation

$$f_n \equiv f(t_n, x_n) \tag{2.9}$$

for the value of the derivative at $t = t_n$, so that Euler's method may be written as

$$x_{n+1} = x_n + h f_n. \tag{2.10}$$

On occasions (such as when dealing with components of systems of ODEs) it is more convenient to write x'_n for the approximation of the derivative of x at $t = t_n$. Then Euler's method would be written as

$$x_{n+1} = x_n + h x'_n. \tag{2.11}$$

---

[1] $\lfloor x \rfloor$ is the "floor" function—take the integer part of x and ignore the fractional part.

Example 2.2

Use Euler's method to solve the IVP

$$x'(t) = 2x(t)(1 - x(t)), \quad t > 10,$$
$$x(10) = 1/5,$$

for $10 \leq t \leq 11$ with $h = 0.2$.

With $f(t, x) = 2x(1 - x)$ and $\eta = 1/5$, we calculate, for each $n = 0, 1, 2, \ldots$, the values[2] t_n, x_n and x'_n. The calculations for the first five steps are shown in Table 2.2 and the points $\{(t_n, x_n)\}$, when the computation is extended to the interval $10 \leq t \leq 13$, are shown in Figure 2.3 by dots; the solid curve shows the exact solution of the IVP: $x(t) = 1/(1 + 4\exp(2(10 - t)))$.   □

n	t_n	x_n	$x'_n = 2x_n(1 - x_n)$
0	10.0	0.2	0.32 (starting condition)
1	10.2	0.2640	0.3886
2	10.4	0.3417	0.4499
3	10.6	0.4317	0.4907
4	10.8	0.5298	0.4982
5	11.0	0.6295	0.4665

Table 2.2 Numerical results for Example 2.2 for $10 \leq t \leq 11$ with $h = 0.2$

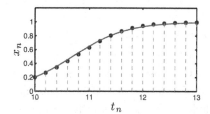

Fig. 2.3 Numerical solution to Example 2.2 (dots) and the exact solution of the IVP (solid curve)

2.4 Analysing the Method

In this section the behaviour of numerical methods, and Euler's method in particular, is investigated in the limit $h \to 0$. When h is decreased, it is required

---

[2]It is not necessary to tabulate the values of t_n in this example since the ODE is autonomous—its right-hand side does not involve t explicitly. It is, nevertheless, good practice to record its value.

that the numerical solution should approach the exact solution, i.e. the size of the GE

$$|e_n| \equiv |x(t_n) - x_n|$$

at $t = t_0 + nh$ should also decrease. This is intuitively reasonable; as we put in more computational effort, we should obtain a more accurate solution. The situation is, however, a little more subtle than is immediately apparent. If we were to consistently compare, say, the fourth terms in the sequences $\{x(t_n)\}$ and $\{x_n\}$ computed with $h = 0.5, 0.25$, and 0.125, then we would compute the error at $t_4 = t_0 + 2.0, t_0 + 1.0$ and $t_0 + 0.5$, respectively. That is, we would be comparing errors at different times when different values of h were employed. Even more worryingly, as $h \to 0$, $t_4 = t_0 + 4h \to t_0$, and we would eventually be comparing x_4 with $x(t_0) = \eta$—the initial condition. What must be done is to compare the exact solution of the IVP and the numerical solution at a fixed time $t = t^*$, say, within the interval of integration. For this, the relevant value of the index n is calculated from $t_n = t_0 + nh = t^*$, or $n = (t^* - t_0)/h$, so that $n \to \infty$ as $h \to 0$. The situation is illustrated in Figure 2.4.

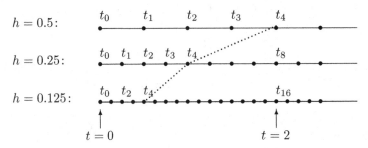

Fig. 2.4 The grids associated with grid sizes $h = 0.5, 0.25$, and 0.125

Some results from Euler's method were analysed in Section 2.1.1 and it appeared that the GE was proportional to h: $e_n \propto h$ when $nh = 0.9$, suggesting that the GE at $t^* = 0.9$ could be made arbitrarily small by choosing a correspondingly small step size h. That is, we could, were we prepared to take a sufficient number of small steps, obtain an approximation that was as accurate as we pleased. This suggests that the Euler's method is *convergent*.

Definition 2.3

A numerical method is said to converge to the solution $x(t)$ of a given IVP at $t = t^*$ if the GE $e_n = x(t_n) - x_n$ at $t_n = t^*$ satisfies

$$|e_n| \to 0 \tag{2.12}$$

as $h \to 0$. It converges at a pth-order rate if $e_n = \mathcal{O}(h^p)$ for some $p > 0$.[3]

We will take the view that numerical methods are of no value unless they are convergent—so any desired accuracy can be guaranteed by taking h to be sufficiently small.

It may be proved that Euler's method converges for IVPs of the form given by Equation (2.1) whenever it has a unique solution for $t_0 \leq t \leq t_f$. However, the proof (see Hairer et al. [28] or Braun [5]) is rather technical owing to the fact that it has to cope with a general nonlinear ODE. We shall therefore be less ambitious and provide a proof only for a linear constant-coefficient case; the conclusions we draw will be relevant to more general situations. In particular, the result will indicate how the local errors committed at each step by truncating a Taylor series accumulate to produce the GE.

Theorem 2.4

Euler's method applied to the IVP

$$
\begin{aligned}
x'(t) &= \lambda x(t) + g(t), \quad 0 < t \leq t_f, \\
x(0) &= 1,
\end{aligned}
$$

where $\lambda \in \mathbb{C}$ and g is a continuously differentiable function, converges and the GE at any $t \in [0, t_f]$ is $\mathcal{O}(h)$.

Proof

Euler's method for this IVP gives

$$
x_{n+1} = x_n + \lambda h x_n + h g(t_n) = (1 + \lambda h) x_n + h g(t_n) \tag{2.13}
$$

while, from the Taylor series expansion (2.5) of the exact solution,

$$
\begin{aligned}
x(t_{n+1}) &= x(t_n) + h x'(t_n) + R_1(t_n), \\
&= x(t_n) + h\big(\lambda x(t_n) + g(t_n)\big) + R_1(t_n). \tag{2.14}
\end{aligned}
$$

By subtracting (2.13) from (2.14) we find that the GE $e_n = x(t_n) - x_n$ satisfies the difference equation

$$
e_{n+1} = (1 + h\lambda)e_n + T_{n+1}, \tag{2.15}
$$

where we have written T_{n+1} instead of $R_1(t_n)$ to simplify the notation. Furthermore, since $x_0 = x(t_0) = \eta$, we have $e_0 = 0$. Equation (2.15) dictates how

---

[3]By convention we refer to the largest such value of p as the "order of the method."

the GE at the next step (e_{n+1}) combines the LTE committed at the current step (T_{n+1}) with the GE inherited from earlier steps (e_n). A similar equation holds for more general ODEs although λ would have to be allowed to vary from step to step.

Substituting $n = 0, 1, 2$ into (2.15) we find, using $e_0 = 0$,

$$e_1 = T_1,$$
$$e_2 = (1 + h\lambda)e_1 + T_2 = (1 + h\lambda)T_1 + T_2,$$
$$e_3 = (1 + h\lambda)e_2 + T_3 = (1 + h\lambda)^2 T_1 + (1 + h\lambda)T_2 + T_3,$$

which suggests the general formula[4]

$$e_n = (1 + h\lambda)^{n-1}T_1 + (1 + h\lambda)^{n-2}T_2 + \cdots + T_n$$
$$= \sum_{j=1}^{n}(1 + h\lambda)^{n-j}T_j. \tag{2.16}$$

All that remains is to find an upper bound for the right-hand side. First, using Exercise 2.8 (with $x = |\lambda|$),

$$|1 + h\lambda| \leq 1 + h|\lambda| \leq e^{h|\lambda|}$$

and so

$$|1 + h\lambda|^{n-j} \leq e^{(n-j)h|\lambda|} = e^{|\lambda|t_{n-j}} \leq e^{|\lambda|,t_f}$$

since $(n - j)h = t_{n-j} \leq t_f$ for $nh \leq t_f$ and $0 < j \leq n$.

Second, since $|T_j| \leq C h^2$ for some constant C (independent of h or j), each term in the summation on the right of (2.16) is bounded by $h^2 C e^{|\lambda|t_f}$ and so

$$|e_n| \leq nh^2 C e^{|\lambda|t_f} = ht_f C e^{|\lambda|t_f}$$

(using $nh = t_f$). Thus, so long as t_f is finite, $e_n = \mathcal{O}(h)$ and we have proved that Euler's method converges at a first-order rate. $\square$

The proof makes it clear that the contribution of the LTE T_j at time $t = t_j$ to the approximation of x_n at time $t = t_n$ is $(1 + h\lambda)^{n-j}T_j$, with $1 + h\lambda > -1$. The LTE is amplified if λ is real and positive, and diminished if λ is real and negative. The most important observation, however, is that an LTE $T_n = \mathcal{O}(h^{p+1})$ of order $(p + 1)$ leads to a GE $e_n = \mathcal{O}(h^p)$ of order p; the cumulative effect of introducing a truncation error at each step is to lose one power of h.

---

[4]This can be proved to be correct either by the process described in Exercise 2.9 or by induction.

2.5 Application to Systems

We shall illustrate by an example how Euler's method applies to systems of ODEs.

Example 2.5

Use Euler's method to compute an approximate solution at $t = 0.2$ of the IVP $x''(t) + x(t) = t$, $t > 0$, with $x(0) = 1$ and $x'(0) = 2$. Use a step length $h = 0.1$.

In order to convert the second-order equation to a system, let $u = x$ and $v = x'$, so $v' = u'' = x'' = -u + t$. This gives the system

$$\left. \begin{array}{rl} u'(t) &= v(t) \\ v'(t) &= t - u(t) \end{array} \right\} \tag{2.17}$$

on the interval $t > 0$ with initial conditions $u(0) = 1, v(0) = 2$. By Taylor series,

$$u(t + h) = u(t) + hu'(t) + \mathcal{O}(h^2),$$
$$v(t + h) = v(t) + hv'(t) + \mathcal{O}(h^2).$$

Neglecting the remainder terms gives Euler's method for the system (2.17):

$$t_{n+1} = t_n + h,$$
$$u_{n+1} = u_n + hu'_n, \qquad u'_{n+1} = v_{n+1},$$
$$v_{n+1} = v_n + hv'_n, \qquad v'_{n+1} = t_{n+1} - u_{n+1}.$$

Note that both u_{n+1} and v_{n+1} must generally be calculated before calculating the derivative approximations u'_{n+1} and v'_{n+1}. The starting conditions are $u_0 = 1$, $v_0 = 2$ at $t = t_0 = 0$ and the given differential equations lead to $u'_0 = v_0 = 2$ and $v'_0 = t_0 - u_0 = -1$. Applying the above recurrence relations first with $n = 0$ and then $n = 1$ gives

$$n = 0: \quad t_1 = 0.1, \qquad\qquad n = 1: \quad t_2 = 0.2,$$
$$u_1 = u_0 + 0.1u'_0 = 1.2, \qquad\qquad u_2 = 1.39,$$
$$v_1 = v_0 + 0.1v'_0 = 1.9, \qquad\qquad v_2 = 1.79,$$
$$u'_1 = 1.9, \qquad\qquad u'_2 = 1.79,$$
$$v'_1 = t_1 - u_1 = -1.1, \qquad\qquad v'_2 = t_2 - u_2 = -1.19.$$

The computations proceed in a similar fashion until the required end time is reached.   □

EXERCISES

2.1.* Use Euler's method with $h = 0.2$ to show that the solution of the IVP $x'(t) = t^2 - x(t)^2$, $t > 0$, with $x(0) = 1$ is approximately $x(0.4) \approx 0.68$.

Show that this estimate changes to $x(0.4) \approx 0.708$ if the calculation is repeated with $h = 0.1$.

2.2.** Obtain the recurrence relation that enables x_{n+1} to be calculated from x_n when Euler's method is applied to the IVP $x'(t) = \lambda x(t)$, $x(0) = 1$ with $\lambda = -10$. In each of the cases $h = 1/6$ and $h = 1/12$

(a) calculate x_1, x_2 and x_3,

(b) plot the points (t_0, x_0), (t_1, x_1), (t_2, x_2), and (t_3, x_3) and compare with a sketch of the exact solution $x(t) = e^{\lambda t}$.

Comment on your results. What is the largest value of h that can be used when $\lambda = -10$ to ensure that $x_n > 0$ for all $n = 1, 2, 3, \ldots$?

2.3.** Apply Euler's method to the IVP

$$\left. \begin{aligned} x'(t) &= 1 + t - x(t), \quad t > 0 \\ x(0) &= 0 \end{aligned} \right\}.$$

Calculate $x_1, x_2, \ldots$ and deduce an expression for x_n in terms of $t_n = nh$ and thereby guess the exact solution of the IVP. Use the expression (2.6) to calculate the LTE and then appeal to the proof of Theorem 2.4 to explain why $x_n = x(t_n)$.

2.4.* Derive Euler's method for the first-order system

$$u'(t) = -2u(t) + v(t)$$
$$v'(t) = -u(t) - 2v(t)$$

with initial conditions $u(0) = 1$, $v(0) = 0$. Use $h = 0.1$ to compute approximate values for $u(0.2)$ and $v(0.2)$.

2.5.* Rewrite the IVP

$$x''(t) + x(t)x'(t) + 4x(t) = t^2, \quad t > 0,$$
$$x(0) = 0, \quad x'(0) = 1,$$

as a first-order system and use Euler's method with $h = 0.1$ to estimate $x(0.2)$ and $x'(0.2)$.

2.6.* Derive Euler's method for the first-order systems obtained in Exercise 1.5 (b) and (c).

2.7.*** This question concerns approximations to the IVP

$$x''(t) + 3x'(t) + 2x(t) = t^2, \quad t > 0,$$
$$x(0) = 1, \qquad x'(0) = 0. \tag{2.18}$$

(a) Write the above initial value problem as a first-order system and hence derive Euler's method for computing approximations to $x(t_{n+1})$ and $x'(t_{n+1})$ in terms of approximations to $x(t_n)$ and $x'(t_n)$.

(b) By eliminating y, show that the system

$$\left. \begin{array}{ll} x'(t) & = y(t) - 2x(t) \\ y'(t) & = t^2 - y(t) \end{array} \right\}$$

has the same solution $x(t)$ as the IVP (2.18) provided that $x(0) = 1$, and that $y(0)$ is suitably chosen. What is the appropriate value of $y(0)$?

(c) Apply Euler's method to the system in part (b) and give formulae for computing approximations to $x(t_{n+1})$ and $y(t_{n+1})$ in terms of approximations to $x(t_n)$ and $y(t_n)$.

(d) Show that the approximations to $x(t_2)$ produced by the methods in (a) and (c) are identical provided both methods use the same value of h.

2.8.** Prove that $e^x \geq 1 + x$ for all $x \geq 0$. [Hint: use the fact that $e^t \geq 1$ for all $t \geq 0$ and integrate both sides over the interval $0 \leq t \leq x$ (where $x \geq 0$).]

2.9.** Replace n by $j-1$ in the recurrence relation (2.15) and divide both sides by $(1 + \lambda h)^j$ to obtain

$$\frac{e_j}{(1+\lambda h)^j} - \frac{e_{j-1}}{(1+\lambda h)^{j-1}} = \frac{T_j}{(1+\lambda h)^j}.$$

By summing both sides from $j = 1$ to $j = n$, show that the result simplifies to give Equation (2.16).

3

The Taylor Series Method

3.1 Introduction

Euler's method was introduced in Chapter 2 by truncating the $\mathcal{O}(h^2)$ terms in the Taylor series of $x(t_h + h)$ about the point $t = t_n$. The accuracy of the approximations generated by the method could be controlled by adjusting the step size h—a strategy that is not always practical, since one may need an inordinate number of steps for high accuracy. For instance, around 1 million steps are necessary to solve the IVP of Example 2.1 to an accuracy of about 10^{-6} for $0 \le t \le 1$.

An alternative is to use a more sophisticated recurrence relation at each step in order to achieve greater accuracy (for the same value of h) or a similar level of accuracy with a larger value of h (and, therefore, fewer steps).

There are many ways of attaining this goal. In this chapter we investigate the possibility of improving the efficiency by including further terms in the Taylor series. Other means will be developed in succeeding chapters.

We shall again be concerned with the solution of an IVP of the form

$$\left. \begin{aligned} x'(t) &= f(t, x), \quad t > t_0 \\ x(t_0) &= \eta \end{aligned} \right\} \tag{3.1}$$

over the interval, $t \in [t_0, t_f]$. We describe a second-order method before treating the case of general order p.

D.F. Griffiths, D.J. Higham, *Numerical Methods for Ordinary Differential Equations*,
Springer Undergraduate Mathematics Series, DOI 10.1007/978-0-85729-148-6_3,
© Springer-Verlag London Limited 2010

3.2 An Order-Two Method: TS(2)

As discussed in Appendix B, the second-order Taylor series expansion

$$x(t + h) = x(t) + hx'(t) + \tfrac{1}{2!}h^2x''(t) + R_2(t)$$

has remainder term $R_2(t) = \mathcal{O}(h^3)$. Setting $t = t_n$ we obtain (since $t_{n+1} = t_n + h$)

$$x(t_{n+1}) = x(t_n) + hx'(t_n) + \tfrac{1}{2!}h^2x''(t_n) + \mathcal{O}(h^3).$$

Neglecting the remainder term on the grounds that it is small leads to the formula

$$x_{n+1} = x_n + hx_n' + \tfrac{1}{2}h^2x_n'', \tag{3.2}$$

in which x_n, x_n', and x_n'' denote approximations to $x(t_n)$, $x'(t_n)$, and $x''(t_n)$ respectively. We shall refer to this as the TS(2) method (some authors call it the three-term TS method—in our naming regime, Euler's method becomes TS(1)). As in Chapter 2, the value of x_n' can be computed from the IVP (3.1):

$$x_n' = f(t_n, x_n).$$

For x_n'' we need to differentiate both sides of the ODE, as illustrated on the following example (which was previously used in Example 2.1 for Euler's method).

Example 3.1

Apply the TS(2) method (3.2) to solve the IVP

$$\left.\begin{array}{c} x'(t) = (1 - 2t)x(t), \quad t > 0 \\ x(0) = 1 \end{array}\right\} \tag{3.3}$$

using $h = 0.3$ and $h = 0.15$ and compare the accuracy at $t = 1.2$ with that of Euler's method, given that the exact solution is $x(t) = \exp[\tfrac{1}{4} - (t - \tfrac{1}{2})^2]$.

In order to apply the formula (3.2) we must express $x''(t)$ in terms of $x(t)$ and t (it could also involve $x'(t)$, but this can be substituted for from the ODE): with the chain rule we find (the general case is dealt with in Exercise 3.9)

$$x''(t) = \frac{\mathrm{d}}{\mathrm{d}t}\left[(1 - 2t)x(t)\right] = -2x(t) + (1 - 2t)x'(t)$$
$$= [(1 - 2t)^2 - 2]x(t),$$

and so the TS(2) method is given by

$$x_{n+1} = x_n + h(1 - 2t_n)x_n + \tfrac{1}{2}h^2[(1 - 2t_n)^2 - 2]x_n, \quad n = 0, 1, \ldots,$$

where $t_n = nh$ and $x_0 = 1$. For the purposes of hand calculation it can be preferable to use (3.2) and arrange the results as shown below:

$$n = 0: \quad t_0 = \ 0, \qquad n = 1: \quad t_1 = t_0 + h = 0.3,$$
$$x_0 = \ 1, \qquad\qquad\qquad x_1 = x_0 + hx_0' + \tfrac{1}{2}h^2x_0'' = 1.2550,$$
$$x_0' = \ 1, \qquad\qquad\qquad x_1' = (1 - 2t_1)x_1 = 0.5020,$$
$$x_0'' = -1, \qquad\qquad\qquad x_1'' = [(1 - 2t_1)^2 - 2]x_1 = -2.3092,$$

with a similar layout for $n = 2, 3, \ldots$.

In Figure 3.1 the computations are extended to the interval $0 \le t \le 4$ and the numerical values with associated GEs at $t = 1.2$ are tabulated in Table 3.1. We observed in Example 2.1 that the GE in Euler's method was halved by halving h, reflecting the relationship $e_n \propto h$. However, from Table 3.1, we see that the error for the TS(2) method is reduced by a factor of roughly 4 as h is halved ($0.0031 \approx 0.0118/4$), suggesting that the GE $e_n \propto h^2$.

We deduce from Table 3.1 that, at $t = 1.2$,[1]

$$\text{GE for Euler's method} \ \approx \ -0.77h$$
$$\text{GE for TS(2) method} \ \approx \ \ \ 0.14h^2.$$

These results suggest that, to achieve an accuracy of 0.01, the step size in Euler's method would have to satisfy $0.77h = 0.01$, from which $h \approx 0.013$ and we would need about $1.2/0.013 \approx 92$ steps to integrate to $t = 1.2$. What are the corresponding values for TS(2)? (Answer: $h \approx 0.27$ and five steps). This illustrates the huge potential advantages of using a higher order method. $\square$

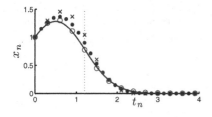

Fig. 3.1 Numerical solutions for Example 3.1

$\times$: Euler's method, $h = 0.3$,

$\bullet$: Euler's method, $h = 0.15$,

$\circ$: TS(2) method, $h = 0.15$.

	Solutions at $t = 1.2$		GEs at $t = 1.2$		
h	Euler: TS(1)	TS(2)	Euler: TS(1)	TS(2)	GE for TS(2)/h^2
0.30	1.0402	0.7748	-0.2535	0.0118	0.131
0.15	0.9014	0.7836	-0.1148	0.0031	0.138

Table 3.1 Numerical solutions and global errors at $t = 1.2$ for Example 3.1. The exact solution is $x(1.2) = e^{-0.24} = 0.7866$.

[1] These relations were deduced from the assertions $e_n = C_1 h$ for Euler and $e_n = C_2 h^2$ for TS(2) and choosing the constants C_1, C_2 so as to match the data in Table 3.1 for $h = 0.15$.

3.2.1 Commentary on the Construction

It was shown in Figure 2.1 that Euler's method could be viewed as the construction of a polygonal curve; the vertices of the polygon representing the numerical values at the points (t_n, x_n) and the gradients of the line segments being dictated by the right-hand side of the ODE evaluated at their left endpoints.

For TS(2) we can take a similar, view with the sides of the "polygon" being quadratic curves given by

$$x = x_n + (t - t_n)x_n' + \tfrac{1}{2}h^2(t - t_n)^2 x_n''$$

for $t_n \leq t \leq t_{n+1}$. These curves are shown as connecting the points P_n and P_{n+1} in Figure 3.2 for $n = 0, 1, 2$ when $h = 0.3$ for the IVP in Example 3.1. These are directly comparable to the polygonal case in Figure 2.1, and the improvement provided by TS(2) is seen to be dramatic.

3.3 An Order-p Method: TS(p)

It is straightforward to extend the TS(2) method described in the previous section to higher order, so we simply sketch the main ideas. The pth- order Taylor series of $x(t + h)$ with remainder is given by

$$x(t + h) = x(t) + hx'(t) + \tfrac{1}{2!}h^2 x''(t) + \cdots + \tfrac{1}{p!}h^p x^{(p)}(t) + R_p(t) \qquad (3.4)$$

as discussed in Appendix B. When $x(t)$ is $(p + 1)$ times continuously differentiable on the interval (t_0, t_f) the remainder term can be written as

$$R_p(t) = \frac{1}{(p + 1)!}h^{p+1}x^{(p+1)}(\xi), \qquad \xi \in (t, t + h),$$

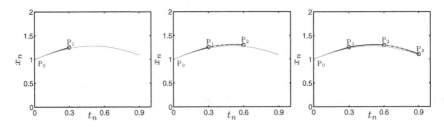

Fig. 3.2 The development of TS(2) for the IVP in Example 3.1 over the first three time steps. The exact solution of the IVP is shown as a solid curve

and, if $|x^{(p+1)}(t)| \leq M$ for all $t \in (t_0, t_f)$, then

$$|R_p(t)| \leq \frac{M}{(p+1)!} h^{p+1},$$

whence $R_p(t) = \mathcal{O}(h^{p+1})$.

The TS(p) method is obtained by applying the expansion at $t = t_n$ and ignoring the remainder term to give

$$x_{n+1} = x_n + hx'_n + \tfrac{1}{2}h^2 x''_n + \cdots + \tfrac{1}{p!} h^p x_n^{(p)}, \qquad (3.5)$$

in which $x_n, x'_n, \ldots x_n^{(p)}$ denote approximations to $x(t_n), x'(t_n) \ldots, x^{(p)}(t_n)$, respectively. It is necessary to differentiate the right side of the ODE $(p-1)$ times in order to complete the specification of the method.

We note that, if the numerical and exact solutions were to coincide at the start of a step, $x_n = x(t_n)$ (this is called the localizing assumption), then the error at the end of the step would be $x(t_{n+1}) - x_{n+1} = R_p(t_n)$. This means that we can interpret the remainder term $R_p(t_n)$, more usually called the LTE, as a measure of the error committed at each step.

3.4 Convergence

We avoid a proof of convergence for general problems and instead generalize Theorem 2.4 given in Section 2.4 for Euler's method.

Theorem 3.2

The Taylor series method TS(p) applied to the IVP

$$x'(t) = \lambda x(t) + g(t), \quad 0 < t \leq t_f,$$
$$x(0) = 1,$$

where $\lambda \in \mathbb{C}$ and g is a p times continuously differentiable function, converges and the GE at any $t \in [0, t_f]$ is $\mathcal{O}(h^p)$.

Proof

See Exercises 3.10 and 3.11.

In summary, the issues that have to be addressed in updating the proof for Euler's method are:

1. showing that the conditions given on g ensure that $x(t)$ is a $(p + 1)$ times continuously differentiable function;

2. determining a suitable modification of equation (2.15);

3. finding an appropriate generalization of Exercise 2.8.

$\square$

3.5 Application to Systems

We illustrate the application of TS(2) to systems of ODEs by solving the same IVP as in Section 2.5.

Example 3.3

Use the TS(2) method with a step length $h = 0.1$ to compute an approximate solution at $t = 0.2$ of the IVP

$$u'(t) = v(t),$$
$$v'(t) = t - u(t) \tag{3.6}$$

on the interval $t > 0$ with initial conditions $u(0) = 1, v(0) = 2$.

The formulae for updating u and v are

$$u_{n+1} = u_n + h u'_n + \tfrac{1}{2} h^2 u''_n,$$
$$v_{n+1} = v_n + h v'_n + \tfrac{1}{2} h^2 v''_n.$$

Since $u'(t) = v(t)$ we have

$$u''(t) = v'(t) = t - u(t).$$

Differentiating $v'(t) = t - u(t)$ leads to

$$v''(t) = 1 - u'(t) = 1 - v(t),$$

which gives us the requisite formulae:

$$u'_n = v_n, \quad u''_n = v'_n = t_n - u_n, \quad v''_n = 1 - v_n.$$

Note that $u'_n = v_n$ and $u''_n = v'_n$ will always be the case when a second-order differential equation is converted to a first-order system. The results of the calculation are laid out in the table below.

n	t_n	u_n	$u'_n = v_n$	$u''_n = v'_n$	v''_n
0	0	1.0000	2.0000	−1.0000	−1.0000
1	0.1	1.1950	1.8950	−1.0950	−0.8950
2	0.2	1.3790	1.7810	−1.1790	−0.7810

The component parts of the recurrence relations can be combined to give

$$u_{n+1} = u_n + hv_n + \tfrac{1}{2}h^2(t_n - u_n),$$
$$v_{n+1} = v_n + h(t_n - u_n) + \tfrac{1}{2}h^2(1 - v_n),$$

but these are perhaps less convenient for computation by hand.

3.6 Postscript

The examples in this chapter show the dramatic improvements in efficiency (measured roughly as the amount of computation needed to attain a specific accuracy) that can be brought about by increasing the order of a method. The Taylor series approach makes this systematic, but each increase in order is accompanied by the need to differentiate the right-hand side of the ODE one more time. If the right hand side of an ODE is given by a complicated formula, or if we have a large system, it may not be practical to attempt even the second-order TS method. For this reason Taylor series methods are not widely used.

In the following chapters we will study methods that achieve order greater than one while avoiding the need to differentiate the differential equation.

EXERCISES

3.1.* Use the TS(2) method to solve the IVP

$$x'(t) = 2x(t)\big(1 - x(t)\big), \quad t > 10,$$
$$x(10) = 1/5,$$

with $h = 0.5$ and compare the accuracy of the solution at $t = 11$ with that of Euler's method using $h = 0.2$ (see Example 2.2).

3.2.** Apply the TS(2) method to the IVP

$$\left.\begin{array}{c} x'(t) = 1 + t - x(t), \quad t > 0 \\ x(0) = 0 \end{array}\right\}$$

and derive a recurrence formula for determining x_{n+1} in terms of x_n, t_n, and h. Use this recurrence to calculate $x_1, x_2, \ldots$ and deduce an expression for x_n in terms of n and h. Show that $x_n = x(t_n)$ for $n = 0, 1, 2, \ldots$. Explain your findings by appealing to the nature of the LTE (remainder term) in this case.

3.3.* Derive the TS(2) method for the first-order systems obtained in Exercise 1.5 (a). Use both TS(1) and TS(2) to determine approximate solutions at $t = 0.2$ using $h = 0.1$.

3.4.** One way of estimating the GE without knowledge of the exact solution is to compute approximate solutions at $t = t_n$ using both TS(p) and TS($p+1$). We will denote these by $x_n^{[p]}$ and $x_n^{[p+1]}$, respectively. The GE in the lower order method is, by definition,

$$e_n = x(t_n) - x_n^{[p]}$$

with $e_n = \mathcal{O}(h^p)$ and, for the higher order method: $x(t_n) - x_n^{[p+1]} = \mathcal{O}(h^{p+1})$. Thus,

$$e_n = x_n^{[p+1]} - x_n^{[p]} + \mathcal{O}(h^{p+1}),$$

from which it follows that the leading term in the GE of the lower order method may be estimated by the difference in the two computed solutions.

Use this process on the data in the first three columns of Table 3.1 and compare with the actual GE for Euler's method given in the fourth column.

3.5.** Apply Euler's method to the IVP $x'(t) = \lambda x(t)$, $x(0) = 1$, with a step size h. Assuming that λ is a real number:

(a) What further condition is required on λ to ensure that the solution $x(t) \to 0$ as $t \to \infty$?

(b) What condition on h then ensures that $|x_n| \to 0$ as $n \to \infty$?

Compare the cases where $\lambda = -1$ and $\lambda = -100$.

What is the corresponding condition if TS(2) is used instead of Euler's method?

3.6.** Write down the TS(3) method for the IVP $x'(t) = \lambda x(t)$, $x(0) = 1$. Repeat for the IVP in Example 3.1.

3.7.*** A rough estimate of the effort required in evaluating a formula may be obtained by counting the number of arithmetic operations

$(+, -, \times, \div)$—this is known as the number of *flops* (floating point operations). For example, the calculation of $3 + 4/5^2$ needs three flops. These may be used as a basis for comparing different methods.

What are the flop counts per step for the TS(p) methods for the IVP in Example 3.1 for $p = 1, 2, 3$?

How many flops do you estimate would be required by TS(1) and TS(2) in Example 3.1 to achieve a GE of 0.01? (Use the data provided in Table 3.1.) Comment on your answer.

3.8.** For the IVP

$$u'(t) = v(t), \qquad v'(t) = -u(t), \quad t > 0,$$
$$u(0) = 1, \qquad v(0) = 0,$$

use the chain rule to differentiate $u^2(t) + v^2(t)$ with respect to t. Hence prove that $u^2(t) + v^2(t) = 1$ for all $t \geq 0$.

Use the TS(2) method to derive a means of computing u_{n+1} and v_{n+1} in terms of u_n and v_n.

Prove that this TS(2) approximation satisfies $u_n^2 + v_n^2 = (1 + \frac{1}{4}h^4)^n$ when $u_0 = 1$ and $v_0 = 0$. (The issue of preserving invariants of an ODE is discussed in Chapter 14.)

3.9.* If $x'(t) = f(t, x(t))$, show that

$$x''(t) = f_t(t, x) + f(t, x)f_x(t, x),$$

where f_t and f_x are the partial derivatives of $f(t, x)$.

3.10.*** Prove that

$$e^x \geq 1 + x + \frac{1}{2!}x^2 + \cdots + \frac{1}{p!}x^p$$

for all $x \geq 0$. [Hint: use induction starting with the case given in Exercise 2.9 and integrate to move from one induction step to the next.]

3.11.*** When $x(t)$ denotes the solution of the IVP in Theorem 3.2, the first p derivatives required for the Taylor expansion (3.4) of $x(t_n + h)$ may be obtained by repeated differentiation of the ODE:

$$x'(t_n) = \lambda x(t_n) + g(t_n),$$
$$x''(t_n) = \lambda x'(t_n) + g'(t_n),$$

$$\vdots$$

$$x^{(p)}(t_n) = \lambda x^{(p-1)}(t_n) + g(t_n).$$

By using analogous relationships between approximations of the derivatives ($x_n^{(j+1)} = \lambda x_n^{(j)} + g^{(j)}(t_n)$) for the $(j+1)$th derivative, for example) show that the GE for the TS(p) method (3.5) for the same IVP satisfies

$$e_{n+1} = r(\lambda h)e_n + T_{n+1}, \qquad n = 0, 1, 2, \ldots,$$

where $r(s) = 1 + s + \frac{1}{2!}s^2 + \cdots + \frac{1}{p!}s^p$, the first $(p+1)$ terms in the Maclaurin expansion of e^s, and $T_{n+1} = \mathcal{O}(h^{p+1})$ [compare with Equation (2.15)].

Hence complete the proof of Theorem 3.2.

4

Linear Multistep Methods—I: Construction and Consistency

4.1 Introduction

The effectiveness of the family of TS(p) methods has been evident in the preceding chapter. For order $p > 1$, however, they suffer a serious disadvantage in that they require the right-hand side of the differential equation to be differentiated a number of times. This often rules out their use in real-world applications, which generally involve (large) systems of ODEs whose differentiation is impractical unless automated tools are used [23]. We look, therefore, for alternatives that do not require the use of second and higher derivatives of the solution.

The families of *linear multistep methods* (LMMs) that we turn to next generally achieve higher order by exploiting the "history" that is available—values of x and x' that were computed at the previous k steps are combined to generate an approximation at the next step. This information is assimilated via what are effectively multipoint Taylor expansions.

We begin with two illustrative examples before describing a more general strategy. In order to distinguish between Taylor expansions of arbitrary functions (that are assumed to have as many continuous derivatives as our expansions require) and Taylor expansions of solutions of our differential equations (which may not have the requisite number of continuous derivatives) we use $z(t)$ for the former.

D.F. Griffiths, D.J. Higham, *Numerical Methods for Ordinary Differential Equations*,
Springer Undergraduate Mathematics Series, DOI 10.1007/978-0-85729-148-6_4,
© Springer-Verlag London Limited 2010

The development of the TS(2) method in Chapter 3 began with the Taylor expansion

$$z(t + h) = z(t) + hz'(t) + \tfrac{1}{2}h^2 z''(t) + \mathcal{O}(h^3) \tag{4.1}$$

and proceeded by applying this with $z = x$, the solution of our IVP, and neglecting the remainder term.[1] At this stage

$$x(t + h) = x(t) + hf(t, x(t)) + \tfrac{1}{2}h^2 x''(t) + \mathcal{O}(h^3).$$

In TS(2) x'' was obtained by differentiating $x'(t) = f(t, x(t))$. LMMs, on the other hand, avoid this by using an approximation to $x''(t)$. There are several possible ways of doing this, of which we will describe two (more systematic derivations of both methods will be given later in this chapter).

4.1.1 The Trapezoidal Rule

We use a second Taylor expansion—that of $z'(t + h)$:

$$z'(t + h) = z'(t) + hz''(t) + \mathcal{O}(h^2), \tag{4.2}$$

so that $hz''(t) = z'(t+h) - z'(t) + \mathcal{O}(h^2)$ (the sign of order terms is immaterial), which, when substituted into (4.1), leads to

$$\begin{aligned} z(t + h) &= z(t) + hz'(t) + \tfrac{1}{2}h\left[z'(t + h) - z'(t) + \mathcal{O}(h^2)\right] + \mathcal{O}(h^3) \\ &= z(t) + \tfrac{1}{2}h\left[z'(t + h) + z'(t)\right] + \mathcal{O}(h^3). \end{aligned} \tag{4.3}$$

It is worth emphasizing that this expansion is valid for *any* three-times continuously differentiable function $z(t)$. We now apply it with $z = x$, the solution of our ODE $x' = f(t, x)$, to give

$$x(t + h) = x(t) + \tfrac{1}{2}h\left[f(t + h, x(t + h)) + f(t, x(t))\right] + \mathcal{O}(h^3). \tag{4.4}$$

Evaluating this at $t = t_n$ and neglecting the remainder term leads to the *trapezoidal rule*:

$$x_{n+1} = x_n + \tfrac{1}{2}h\left[f(t_{n+1}, x_{n+1}) + f(t_n, x_n)\right]. \tag{4.5}$$

---

[1]Note that rearranging (4.1) gives

$$z'(t) = \frac{z(t + h) - z(t)}{h} + \mathcal{O}(h),$$

which is consistent with the definition of a derivative:

$$z'(t) = \lim_{h \to 0} \frac{z(t + h) - z(t)}{h}.$$

Notation. In order to simplify the presentation of LMMs, we shall use the abbreviations

$$f_n = f(t_n, x_n), \quad f_{n+j} = f(t_{n+j}, x_{n+j}), \text{ etc.} \tag{4.6}$$

These are approximations to the gradient of the solution at $t_n = t_0 + nh$ and $t_{n+j} = t_n + jh$. They should not be confused with $f(t_n, x(t_n))$ and $f(t_{n+j}, x(t_{n+j}))$, which are gradients of the exact solution at these times.

Our trapezoidal rule may then be written as

$$x_{n+1} - x_n = \tfrac{1}{2}h(f_{n+1} + f_n). \tag{4.7}$$

Since $f_{n+1} = f(t_{n+1}, x_{n+1})$ we see that x_{n+1} appears also on the right hand side of (4.7) and so, unlike the TS(p) methods, we do not have a direct expression for x_{n+1} in terms of data available from earlier times. Methods having this property are known as *implicit methods*.

4.1.2 The 2-step Adams–Bashforth method: AB(2)

The treatment for the two-step Adams–Bashforth method (which we will refer to as AB(2)) is similar to that for the trapezoidal rule, except that the expansion (4.2) is replaced with

$$z'(t - h) = z'(t) - hz''(t) + \mathcal{O}(h^2). \tag{4.8}$$

It follows that $hz''(t) = z'(t) - z'(t - h) + \mathcal{O}(h^2)$, which, when substituted into (4.1), gives

$$\begin{aligned}
z(t + h) &= z(t) + hz'(t) + \tfrac{1}{2}h\left[z'(t) - z'(t - h) + \mathcal{O}(h^2)\right] + \mathcal{O}(h^3) \\
&= z(t) + \tfrac{1}{2}h\left[3z'(t) - z'(t - h)\right] + \mathcal{O}(h^3).
\end{aligned} \tag{4.9}$$

Now, with $z = x$, the solution of our ODE $x' = f(t, x)$, this gives

$$x(t + h) = x(t) + \tfrac{1}{2}h\left[3f(t, x(t)) - f(t - h, x(t - h))\right] + \mathcal{O}(h^3).$$

Evaluating this at $t = t_n$ and neglecting the remainder term leads to AB(2):

$$x_{n+1} = x_n + \tfrac{1}{2}h(3f_n - f_{n-1}).$$

In this equation the solution x_{n+1} at time t_{n+1} is given in terms of data at the *two* previous time levels: $t = t_n$ and $t = t_{n-1}$. This is an example of a two-step LMM. It is usual to write such methods in a form where the smallest index is n:

$$x_{n+2} = x_{n+1} + \tfrac{1}{2}h(3f_{n+1} - f_n). \tag{4.10}$$

Example 4.1

Apply the TS(2), trapezoidal and AB(2) methods to the IVP $x'(t) = (1 - 2t)x(t)$, $t \in (0, 4]$, $x(0) = 1$ (cf. Examples 2.1 and 3.1) with $h = 0.2$ and $h = 0.1$. Compare the accuracy achieved by the various methods at $t = 1.2$.

The application of TS(2) is described in Example 3.1. With $f(t, x) = (1 - 2t)x$, the trapezoidal rule gives

$$x_{n+1} = x_n + \tfrac{1}{2}h[(1 - 2t_{n+1})x_{n+1} + (1 - 2t_n)x_n],$$

which can be rearranged to read

$$x_{n+1} = \frac{1 + \tfrac{1}{2}h(1 - 2t_n)}{1 - \tfrac{1}{2}h(1 - 2t_{n+1})}x_n \qquad (4.11)$$

for $n = 0, 1, 2, \ldots$ and $x_0 = 1$. When $h = 0.2$ we take $n = 6$ steps to reach $t = 1.2$, at which point we obtain $x_6 = 0.789\,47$. Comparing this with the exact solution $x(1.2) = 0.786\,63$ gives a GE of -0.0028.

For the AB(2) method, we find

$$\begin{aligned} x_{n+2} &= x_{n+1} + \tfrac{1}{2}h[3(1 - 2t_{n+1})x_{n+1} - (1 - 2t_n)x_n] \\ &= \left[1 + \tfrac{3}{2}h(1 - 2t_{n+1})\right]x_{n+1} - \tfrac{1}{2}h(1 - 2t_n)x_n, \end{aligned}$$

which holds for $n \geq 0$. When $n = 0$ we have $x_0 = 1$, $t_0 = 0$, $t_1 = h$, and

$$x_2 = \left[1 + \tfrac{3}{2}h(1 - 2h)\right]x_1 - \tfrac{1}{2}h.$$

It is necessary to use some other method to find the additional starting value x_1 before we can begin. Two obvious possibilities are to use either Euler's method, $x_1 = (1 + h)x_0$ (we label the results ABE), or the value computed by the trapezoidal rule (4.11) with $n = 0$ (leading to ABT).

The calculation of the sequence $\{x_n\}$ is best organized with a start-up phase: first calculate $f_0 = 1$ from the initial data $x = x_0$ at $t = t_0$. The solution at $t = t_1$ is then calculated by Euler's method (for instance):

$$t_1 = t_0 + h = 0.2,$$
$$x_1 = x_0 + hf_0 = 1.2 \text{ (Euler's method)},$$
$$f_1 = (1 - 2t_1)x_1 = 0.72;$$

then, for each $n = 0, 1, 2, \ldots$, x_{n+2} and f_{n+2} are calculated from

$$t_{n+2} = t_{n+1} + h,$$
$$x_{n+2} = x_{n+1} + \tfrac{1}{2}h(3f_{n+1} - f_n),$$
$$f_{n+2} = (1 - 2t_{n+2})x_{n+2},$$

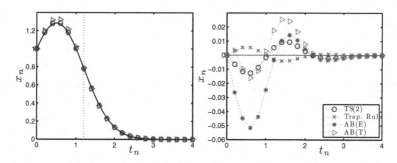

Fig. 4.1 Numerical solutions (left) and GEs (right) as functions of time with $h = 0.2$ for Example 4.1

h	TS(2)	Trap.	ABE	ABT
0.2	5.4	-2.8	-3.6	17.6
0.1	1.4	-0.71	-0.66	4.0
Ratio	3.90	4.00	5.49	4.40

Table 4.1 Global errors (multiplied by 10^3) at $t = 1.2$

so that the value of f at $t = t_{n+2}$ is available in the succeeding steps—i.e. the value of f at any given grid point needs be computed only once.

The solutions obtained with these methods are shown in Figure 4.1 (left) as functions of t_n. These are indistinguishable from the exact solution curve except near to the maximum at $t = \frac{1}{2}$; the corresponding GEs are shown in the right of the figure. The GEs at $t = 1.2$ (obtained by subtracting the numerical solutions from the exact solution) are shown in Table 4.1 for $h = 0.2$ and $h = 0.1$. Dividing the GE when $h = 0.2$ by that when $h = 0.1$ leads to the bottom row in Table 4.1. Since this ratio is close to 4, it suggests that the four methods shown all converge at a second-order rate: $e_n \propto h^2$. Surprisingly, the GE of ABE is *smaller* than that of ABT, which uses a more accurate starting value x_1. However, when we look at the behaviour of the GE over the whole domain in Figure 4.1, we see that ABE has a particularly large GE at $t \approx 0.5$; to sample the errors at only one time may be misleading. Closer scrutiny of the figure reveals that the greatest overall accuracy is given by the trapezoidal rule, and we estimate that

$$\mathrm{GE}_{\mathrm{TS}(2)} \approx -\mathrm{GE}_{\mathrm{trap}}, \quad \mathrm{GE}_{\mathrm{ABT}} \approx -2\,\mathrm{GE}_{\mathrm{trap}}, \quad \mathrm{GE}_{\mathrm{ABE}} \approx -5\,\mathrm{GE}_{\mathrm{trap}}. \quad \square$$

k	p	Method	Name
1	1	$x_{n+1} - x_n = hf_n$	Euler
1	1	$x_{n+1} - x_n = hf_{n+1}$	Backward Euler
1	2	$x_{n+1} - x_n = \frac{1}{2}h(f_{n+1} + f_n)$	trapezoidal
2	2	$x_{n+2} - x_{n+1} = \frac{1}{2}h(3f_{n+1} - f_n)$	two-step Adams–Bashforth
2	2	$x_{n+2} - x_{n+1} = \frac{1}{12}h(5f_{n+2} + 8f_{n+1} - f_n)$	two-step Adams–Moulton
2	4	$x_{n+2} - x_n = \frac{1}{3}h(f_{n+2} + 4f_{n+1} + f_n)$	Simpson's rule
2	3	$x_{n+2} + 4x_{n+1} - 5x_n = h(4f_{n+1} + 2f_n)$	Dahlquist (see Example 4.11)

Table 4.2 Examples of LMMs showing their step number (k) and order (p)

4.2 Two-Step Methods

So far we have encountered three examples of LMMs: Euler's method, the trapezoidal rule (4.7) and the AB(2) method (4.10). The last two were derived from the expansions (4.3) and (4.9), which are valid for all three times differentiable functions z, and which we regard as being generalizations of Taylor series—they relate the values of z and $z'(t)$ at several different points.

For the time being we shall be concerned only with two-step LMMs, such as AB(2), that involve the three time levels t_n, t_{n+1}, and t_{n+2}. For these, we need to find the coefficients α_0, α_1, β_0, β_1, and β_2 so that

$$z(t + 2h) + \alpha_1 z(t + h) + \alpha_0 z(t)$$
$$= h(\beta_2 z'(t + 2h) + \beta_1 z'(t + h) + \beta_0 z'(t)) + \mathcal{O}(h^{p+1}), \quad (4.12)$$

where p might be specified in some cases or we might try to make p as large as possible in others. We have taken $\alpha_2 = 1$ as a normalizing condition (the coefficient of $z(t + 2h)$).

Choosing $z = x$, where $x' = f(t, x)$, and dropping the $\mathcal{O}(h^{p+1})$ remainder term, we arrive at the general two-step LMM

$$x_{n+2} + \alpha_1 x_{n+1} + \alpha_0 x_n = h(\beta_2 f_{n+2} + \beta_1 f_{n+1} + \beta_0 f_n). \quad (4.13)$$

An LMM is said to be *explicit* (of explicit type) if $\beta_2 = 0$ and *implicit* if $\beta_2 \neq 0$. For example, Euler's method ($x_{n+1} = x_n + hf_n$) is an example of an explicit one-step LMM while the trapezoidal rule is an example of an implicit one-step method.

On occasion we may write (4.13) in the form

$$x_{n+2} + \alpha_1 x_{n+1} + \alpha_0 x_n = h(\beta_2 x'_{n+2} + \beta_1 x'_{n+1} + \beta_0 x'_n).$$

Some further examples of one- and two-step LMMs are listed in Table 4.2.

4.2.1 Consistency

In order to streamline the process of determining the coefficients in the LMM (4.13), we introduce the notion of a linear difference operator.

Definition 4.2

The *linear difference operator* $\mathscr{L}_h$ associated with the LMM (4.13) is defined for an arbitrary continuously differentiable function $z(t)$ by

$$\mathscr{L}_h z(t) = z(t + 2h) + \alpha_1 z(t + h) + \alpha_0 z(t) - $$
$$h(\beta_2 z'(t + 2h) + \beta_1 z'(t + h) + \beta_0 z'(t)).$$

Apart from the remainder term, this is the difference between the left- and right-hand sides of (4.12). $\mathscr{L}_h$ is a *linear* operator since, for constants a and b,

$$\mathscr{L}_h(az(t) + bw(t)) = a\mathscr{L}_h z(t) + b\mathscr{L}_h w(t).$$

The construction of new methods amounts to finding suitable coefficients $\{\alpha_j, \beta_j\}$. We shall prove later in this chapter that the coefficients should be determined so as to ensure that the resulting LMM is consistent.

Definition 4.3

A linear difference operator $\mathscr{L}_h$ is said to be *consistent of order* p if

$$\mathscr{L}_h z(t) = \mathcal{O}(h^{p+1})$$

with $p > 0$ for every smooth function z.

An LMM whose difference operator is consistent of order p for some $p > 0$ is said to be *consistent*. A method that fails to meet this requirement is called *inconsistent* (see Exercise 4.9). This definition is in keeping with our findings for TS(p) methods: an LTE of order $(p + 1)$ gives rise to convergence of order p (Theorem 3.2).

Example 4.4

Show that Euler's method is consistent.

The linear difference operator for Euler's method is

$$\mathscr{L}_h z(t) = z(t + h) - z(t) - hz'(t),$$

which, by Taylor expansion, gives

$$\mathscr{L}_h z(t) = \tfrac{1}{2!} h^2 z''(t) + \mathcal{O}(h^3),$$

and so $\mathscr{L}_h z(t) = \mathcal{O}(h^2)$ and the method is consistent of order 1 ($p = 1$). $\square$

Example 4.5

What is the order of consistency of the last method listed in Table 4.2?

The associated linear difference operator is

$$\mathscr{L}_h z(t) = z(t + 2h) + 4z(t + h) - 5z(t) - h(4z'(t + h) + 2z'(t)).$$

With the aid of the Taylor expansions

$$z(t + 2h) = z(t) + 2hz'(t) + 2h^2 z''(t) + \tfrac{4}{3}h^3 z'''(t) + \tfrac{2}{3}h^4 z''''(t) + \mathcal{O}(h^5),$$
$$z(t + h) = z(t) + hz'(t) + \tfrac{1}{2}h^2 z''(t) + \tfrac{1}{6}h^3 z'''(t) + \tfrac{1}{24}h^4 z''''(t) + \mathcal{O}(h^5),$$
$$z'(t + h) = z'(t) + hz''(t) + \tfrac{1}{2}h^2 z'''(t) + \tfrac{1}{6}h^3 z''''(t) + \mathcal{O}(h^4),$$

we find (collecting terms appropriately)

$$\begin{aligned}
\mathscr{L}_h z(t) = {}& [1 + 4 - 5]\, z(t) \\
& + h\,[2 + 4 - [4 + 2]]\, z'(t) \\
& + h^2\,[2 + 2 - 4]\, z''(t) \\
& + h^3\,\left[\tfrac{4}{3} + 4 \times \tfrac{1}{6} - 4 \times \tfrac{1}{2}\right] z'''(t) \\
& + h^4\,\left[\tfrac{2}{3} + 4 \times \tfrac{1}{24} - 4 \times \tfrac{1}{6}\right] z''''(t) + \mathcal{O}(h^5).
\end{aligned}$$

Thus, $\mathscr{L}_h z(t) = \tfrac{1}{6} h^4 z''''(t) + \mathcal{O}(h^5)$ and so $\mathscr{L}_h z(t) = \mathcal{O}(h^4)$ and the method is consistent of order $p = 3$. It is, in fact, the explicit two-step method having highest possible order. $\square$

4.2.2 Construction

In the previous section the coefficients of LMMs were given and we were then able to find their order and error constants. We now describe how the coefficients may be determined by the *method of undetermined coefficients*.

For the general two-step LMM given by Equation (4.13) the associated linear difference operator is (see Definition 4.2)

$$\mathscr{L}_h z(t) = z(t + 2h) + \alpha_1 z(t + h) + \alpha_0 z(t) - $$
$$h(\beta_2 z'(t + 2h) + \beta_1 z'(t + h) + \beta_0 z'(t)) \quad (4.14)$$

and the right-hand side may be expanded with the aid of the Taylor expansions

$$z(t + 2h) = z(t) + 2hz'(t) + 2h^2 z''(t) + \tfrac{4}{3}h^3 z'''(t) + \ldots,$$
$$z(t + h) = z(t) + hz'(t) + \tfrac{1}{2}h^2 z''(t) + \tfrac{1}{6}h^3 z'''(t) + \ldots,$$
$$z'(t + 2h) = z'(t) + 2hz''(t) + 2h^2 z'''(t) + \tfrac{4}{3}h^3 z''''(t) + \ldots,$$
$$z'(t + h) = z'(t) + hz''(t) + \tfrac{1}{2}h^2 z'''(t) + \tfrac{1}{6}h^3 z''''(t) + \ldots.$$

The precise number of terms that should be retained depends on either the order required[2] or the maximum order possible with the "template" used—some coefficients in the LMM may be set to zero in order to achieve a method with a particular pattern of terms. It will also become clear as we proceed that it is generally advantageous not to fix all the coefficients so as to achieve maximum order of consistency, but to retain some free parameters to meet other demands (notably stability).

We focus initially on the issue of *consistency*, i.e. what is needed for methods to have order at least $p = 1$. Expanding the right of (4.14) we find (collecting terms appropriately)

$$\mathscr{L}_h z(t) = \left(1 + \alpha_1 + \alpha_0\right)z(t) + h\left[2 + \alpha_1 - (\beta_2 + \beta_1 + \beta_0)\right]z'(t) + \mathcal{O}(h^2).$$

We shall have

$$\mathscr{L}_h z(t) = \mathcal{O}(h^2)$$

and, therefore, consistency of order 1, if the coefficients are chosen so that

$$
\begin{aligned}
1 + \alpha_1 + \alpha_0 &= 0, \\
2 + \alpha_1 &= \beta_2 + \beta_1 + \beta_0.
\end{aligned}
\tag{4.15}
$$

These conditions can be written more concisely if we introduce two polynomials.

Definition 4.6

The *first* and *second characteristic polynomials* of the LMM

$$x_{n+2} + \alpha_1 x_{n+1} + \alpha_0 x_n = h(\beta_2 f_{n+2} + \beta_1 f_{n+1} + \beta_0 f_n)$$

are defined to be

$$\rho(r) = r^2 + \alpha_1 r + \alpha_0, \quad \sigma(r) = \beta_2 r^2 + \beta_1 r + \beta_0 \tag{4.16}$$

respectively.

---

[2] For order p all terms up to those containing h^{p+1} must be retained but, since the β-terms are already multiplied by h, the expansions of the z' terms need only include terms up to h^p.

The following result is a direct consequence of Definition 4.3 and writing conditions (4.15) in terms of the characteristic polynomials.

Theorem 4.7

The two-step LMM

$$x_{n+2} + \alpha_1 x_{n+1} + \alpha_0 x_n = h(\beta_2 f_{n+2} + \beta_1 f_{n+1} + \beta_0 f_n)$$

is *consistent* with the ODE $x'(t) = f(t, x(t))$ if, and only if,

$$\rho(1) = 0 \quad \text{and} \quad \rho'(1) = \sigma(1).$$

In general, when the right side of the linear difference operator (4.14) is expanded to higher order terms and these are collected appropriately, it is found that

$$\mathscr{L}_h z(t) = C_0 z(t) + C_1 h z'(t) + \cdots + C_p h^p z^{(p)}(t) + \mathcal{O}(h^{p+1}), \qquad (4.17)$$

where, as in the lead up to (4.15), $C_0 = 1 + \alpha_1 + \alpha_0$ and $C_1 = 2 + \alpha_1 - (\beta_2 + \beta_1 + \beta_0)$. The coefficients C_j are each linear combinations of the α and β values that do not involve h. In view of Definition 4.3, an LMM will be consistent of order p if

$$C_0 = C_1 = \cdots = C_p = 0,$$

in which case

$$\mathscr{L}_h z(t) = C_{p+1} h^{p+1} z^{(p+1)}(t) + \mathcal{O}(h^{p+2}). \qquad (4.18)$$

The first non-zero coefficient C_{p+1} is known as the *error constant*.

The next theorem sheds light on the significance of consistency.

Theorem 4.8

A convergent LMM is consistent.

Thus consistency is necessary for convergence, but it is *not* true to say that consistent methods are always convergent.

Proof

Suppose that the LMM (4.13) is convergent. Definition 2.3 then implies that $x_{n+2} \to x(t^* + 2h)$, $x_{n+1} \to x(t^* + h)$ and $x_n \to x(t^*)$ as $h \to 0$ when $t_n = t^*$. However, since $t_{n+2}, t_{n+1} \to t^*$, taking the limit on both sides of

$$x_{n+2} + \alpha_1 x_{n+1} + \alpha_0 x_n = h(\beta_2 f_{n+2} + \beta_1 f_{n+1} + \beta_0 f_n)$$

leads to

$$\rho(1)x(t^*) = 0.$$

But $x(t^*) \neq 0$, in general, and so $\rho(1) = 0$, the first of the consistency conditions in Theorem 4.7.

For the second part of the proof the limit $h \to 0$ is taken of both sides of

$$\frac{x_{n+2} + \alpha_1 x_{n+1} + \alpha_0 x_n}{h} = \beta_2 f_{n+2} + \beta_1 f_{n+1} + \beta_0 f_n.$$

The right-hand side converges to the limit $\sigma(1)f(t^*, x(t^*))$ and, for the left-hand side, we use the limits $x_{n+2} \to x(t^* + 2h)$, $x_{n+1} \to x(t^* + h)$, and $x_n \to x(t^*)$ together with l'Hôpital's rule to conclude that

$$\lim_{h \to 0} \frac{x_{n+2} + \alpha_1 x_{n+1} + \alpha_0 x_n}{h} = (2 + \alpha_1)x'(t^*).$$

Thus, the limiting function $x(t)$ satisfies[3]

$$\rho'(1)x'(t^*) = \sigma(1)f(t^*, x(t^*))$$

at $t = t^*$, which is not the correct differential equation unless $\rho'(1) = \sigma(1)$. □

Example 4.9

Determine the coefficients in the 1-step LMM

$$x_{n+1} + \alpha_0 x_n = h(\beta_1 f_{n+1} + \beta_0 f_n)$$

so that the resulting method has order 1. Find the error constant and show that there is a unique method having order 2. What is the error constant for the resulting method?

The associated linear difference operator is, by definition,

$$\mathscr{L}_h z(t) = z(t + h) + \alpha_0 z(t) - h(\beta_1 z'(t + h) + \beta_0 z'(t)) \tag{4.19}$$

and Taylor expanding the terms $z(t + h)$ and $z'(t + h)$ gives

$$\mathscr{L}_h z(t) = (1 + \alpha_0)z(t) + \big(1 - (\beta_1 + \beta_0)\big)hz'(t) + \mathcal{O}(h^2).$$

Therefore, we shall have consistency (i.e., order at least 1) if the terms in $z(t)$ and $hz'(t)$ vanish. This will be the case if

$$1 + \alpha_0 = 0 \quad \text{and} \quad 1 = \beta_1 + \beta_0.$$

---

[3] The details are left to Exercise 4.17.

These are two equations in three unknowns and their general solution may be expressed as $\alpha_0 = -1$, $\beta_1 = \theta$, $\beta_0 = 1 - \theta$, which gives rise to the *one-parameter family* of LMMs known as the θ-method:

$$x_{n+1} - x_n = h(\theta f_{n+1} + (1 - \theta)f_n). \tag{4.20}$$

The error constant is C_2, for which we have to retain one more term in each of the Taylor expansions of $z(t + h)$ and $z'(t + h)$. We then find

$$\mathscr{L}_h z(t) = (\tfrac{1}{2} - \theta)h^2 z''(t) + \mathcal{O}(h^3)$$

so that $C_2 = \tfrac{1}{2} - \theta$. The common choices for θ are:

1. $\theta = 0$. Euler's method: $x_{n+1} - x_n = hf_n$.

2. $\theta = 1$. Backward Euler method: $x_{n+1} - x_n = hf_{n+1}$ (see Table 4.2). This is also known as the implicit Euler method.

3. $\theta = \tfrac{1}{2}$. trapezoidal rule: $x_{n+1} - x_n = \tfrac{1}{2}h(f_n + f_{n+1})$. This is the unique value of θ for which $C_2 = 0$ and the method becomes of second order. To compute its error constant C_3, the Taylor expansions must be extended by one term (with $\theta = \tfrac{1}{2}$), so:

$$\begin{aligned} \mathscr{L}_h z(t) = \quad & h^3(\tfrac{1}{6} - \tfrac{1}{2}\theta)z'''(t) + \mathcal{O}(h^4) \\ = \; & -\tfrac{1}{12}h^3 z'''(t) + \mathcal{O}(h^4) \end{aligned}$$

and, therefore, $C_3 = -\tfrac{1}{12}$.   □

Example 4.10

Determine the coefficients in the two-step LMM

$$x_{n+2} + \alpha_0 x_n = h(\beta_1 f_{n+1} + \beta_0 f_n)$$

so that it has as high an order of consistency as possible. What is this order and what is the error constant for the resulting method?

The associated linear difference operator is

$$\mathscr{L}_h z(t) = z(t + 2h) + \alpha_0 z(t) - h(\beta_1 z'(t + h) + \beta_0 z'(t)).$$

The LMM contains three arbitrary constants so we expect to be able to satisfy three linear equations; that is, we should be able to make the terms in $z(t)$, $hz'(t)$, and $h^2 z''(t)$ in the Taylor expansion of $\mathscr{L}_h z(t)$ all vanish leaving $\mathscr{L}_h z(t) = C_3 h^3 z'''(t) + \mathcal{O}(h^4)$. We therefore expect to have order $p = 2$; for this to occur we have to retain terms up to $h^3 z'''(t)$. (For some methods we get a "bonus" in that the next term is automatically zero—see Simpson's method.)

With the aid of the Taylor series

$$z(t + 2h) = z(t) + 2hz'(t) + 2h^2 z''(t) + \tfrac{4}{3}h^3 z'''(t) + \tfrac{2}{3}h^4 z''''(t) + \mathcal{O}(h^5)$$
$$z'(t + h) = \qquad z'(t) + hz''(t) + \tfrac{1}{2}h^2 z'''(t) + \tfrac{1}{6}h^3 z''''(t) + \mathcal{O}(h^4)$$

we find, on collecting terms in powers of h,

$$\mathscr{L}_h z(t) = [1 + \alpha_0] \, z(t) + h \, [2 - (\beta_1 + \beta_0)] \, z'(t)$$
$$+ h^2 \, [2 - \beta_1] \, z''(t)$$
$$+ h^3 \, [\tfrac{4}{3} - \tfrac{1}{2}\beta_1] \, z'''(t) + \mathcal{O}(h^4).$$

Setting the coefficients of the first three terms to zero gives

$$1 + \alpha_0 = 0, \quad 2 = \beta_1 + \beta_0, \quad 2 = \beta_1$$

whose solution is $\alpha_0 = -1$, $\beta_1 = 2$, $\beta_0 = 0$ and the resulting LMM is

$$x_{n+2} - x_n = 2hf_{n+1}, \tag{4.21}$$

known in some circles as the mid-point rule and in others as the "leap-frog" method; we shall see in the next chapter that it belongs to the class of Nyström methods. It follows that

$$\mathscr{L}_h z(t) = \tfrac{1}{3}h^3 z'''(t) + \mathcal{O}(h^4)$$

and so $\mathscr{L}_h z(t) = \mathcal{O}(h^3)$ and the method is consistent of order $p = 2$ with error constant $C_3 = \tfrac{1}{3}$. $\qquad\square$

An alternative approach to constructing LMMs based on interpolating polynomials is suggested by Exercises 4.18 and 4.19.

Is this all there is to constructing LMMs? Theorem 4.8 did *not* say that consistent methods are convergent, and the following numerical example indicates that some further property is required to generate a useful (convergent) method.

Example 4.11

Use the method

$$x_{n+2} + 4x_{n+1} - 5x_n = h(4f_{n+1} + 2f_n)$$

(see Example 4.5) to solve the IVP $x'(t) = -x(t)$ for $t > 0$ with $x(0) = 1$. Use the three different grid sizes $h = 0.1, 0.001$ and 0.0001 and, for the additional starting value, use $x_1 = e^{-h}$.

$h = 0.1$	$h = 0.01$	$h = 0.001$
$x_7 = \quad 0.544$	$x_{13} = \quad 0.938$	$x_{19} = \quad 1.070$
$x_8 = \quad 0.199$	$x_{14} = \quad 0.567$	$x_{20} = \quad 0.535$
$x_9 = \quad 1.735$	$x_{15} = \quad 2.384$	$x_{21} = \quad 3.205$
$x_{10} = -6.677$	$x_{16} = -6.810$	$x_{22} = -10.159$
$x_{11} = \quad 37.706$	$x_{17} = \quad 39.382$	$x_{23} = \quad 56.697$
$x_{12} = -197.958$	$x_{18} = -193.017$	$x_{24} = -277.788$

Table 4.3   The numerical solutions for Example 4.11

Since $f_n = -x_n$ and $f_{n+1} = -x_{n+1}$ the LMM leads to the difference equation

$$x_{n+2} = -4(1 + h)x_{n+1} + (5 - 2h)x_n, \quad n = 0, 1, 2, \ldots,$$

which is used to compute the values given in Table 4.3. In each case we see that the numerical solution oscillates wildly within just a few steps, making the method completely worthless.   □

Behaviour of the type seen in the preceding example had been experienced frequently ever since LMMs were first introduced in the 19th century, but it took until 1956 for the Swedish numerical analyst Germund Dahlquist to discover the cause—this example was used in his original work and reappears in his book with A. Björck [16]. He proved that for a method to be convergent (and therefore useful) it not only had to be consistent (have order ≥ 1), but it also had to satisfy a property known as *zero-stability*. This is addressed in the next chapter.

We conclude this chapter by extending two-step LMMs so as to incorporate further terms from the history of the numerical solution. This generalization will allow us the opportunity of summarizing the main ideas to date.

4.3 k-Step Methods

The one- and two-step methods that have been discussed thus far generalize quite naturally to k-step methods. The most general method takes the form

$$x_{n+k} + \alpha_{k-1}x_{n+k-1} + \cdots + \alpha_0 x_n = h\big(\beta_k f_{n+k} + \beta_{k-1}f_{n+k-1} + \cdots + \beta_0 f_n\big) \quad (4.22)$$

and is of implicit type unless $\beta_k = 0$, when it becomes explicit. It uses k past values of the pair x_{n+j}, f_{n+j} $(j = 0 : k-1)$, as well as f_{n+k} in the implicit case, in order to calculate x_{n+k}. The coefficients have been normalized so that the coefficient of x_{n+k} is $\alpha_k = 1$. This method has first and second characteristic polynomials (see Definition 4.6)

$$\begin{aligned}
\rho(r) &= \quad r^k + \alpha_{k-1}r^{k-1} + \cdots + \alpha_0, \\
\sigma(r) &= \beta_k r^k + \beta_{k-1}r^{k-1} + \cdots + \beta_0,
\end{aligned} \tag{4.23}$$

and an associated linear difference operator defined by (see Definition 4.2)

$$\mathscr{L}_h z(t) \equiv \sum_{j=0}^{k} \alpha_j z(t + jh) - h\beta_j z'(t + jh). \tag{4.24}$$

Taylor expanding the right-hand side about the point $h = 0$ and collecting terms in powers of h leads to (see (4.17))

$$\mathscr{L}_h z(t) = C_0 z(t) + C_1 h z'(t) + \cdots + C_p h^p z^{(p)}(t) + C_{p+1} h^{p+1} z^{(p+1)}(t) + \mathcal{O}(h^{p+2}),$$

in which $C_0 = \rho(1)$, $C_1 = \rho'(1) - \sigma(1)$, and the coefficients $C_0, C_1, C_2, \ldots$ are linear combinations of $\beta_0, \beta_1, \ldots, \beta_k, \alpha_0, \alpha_2, \ldots, \alpha_k$ ($\alpha_k = 1$). The method has order p if

$$C_0 = C_1 = \cdots = C_p = 0$$

and the first non-zero coefficient C_{p+1} is the error constant.[4]

The general implicit (explicit) k-step LMM has $2k + 1$ ($2k$) arbitrary coefficients; so, provided the linear relationships are linearly independent, we would expect to be able to achieve order $2k$ with implicit methods and $(2k - 1)$ with explicit methods. We do not offer proofs of these assertions since it will turn out in the next chapter that, because of the type of instability observed in Example 4.11, convergent methods cannot, in general, achieve such high orders.

EXERCISES

4.1.* Distinguish the implicit methods from the explicit methods in Table 4.2.

4.2.** Consider the problem of using the trapezoidal rule to solve the IVP $x'(t) = -x^2(t)$, $x(0) = 1$. Show that it is necessary to solve a quadratic equation in order to determine x_{n+1} from x_n and that an appropriate root can be identified by making use of the property that $x_{n+1} \to x_n$ as $h \to 0$.

Hence find an approximate solution at $t = 0.2$ using $h = 0.1$.

---

[4]It is common to define the error constant to be $C_{p+1}/\sigma(1)$—we shall call this the *scaled error constant*. See the footnote on page 69 for an explanation.

4.3.** Consider the IVP $x'(t) = 1 + x^2(t)$, $x(0) = 0$. When the backward Euler method is applied to this problem the numerical solution x_1 at time $t_1 = h$ is defined implicitly as the solution of a quadratic equation. Explain carefully why it is appropriate to choose the root given by

$$x_1 = \frac{2h}{1 + \sqrt{1 - 4h^2}}.$$

4.4.** The backward Euler method is to be applied to the IVP $x'(t) = 2\sqrt{x(t)}$, with $x(0) = 1$. Use an argument similar to that in Exercise 4.2 to explain carefully why this leads to

$$x_{n+1} = \left(h + \sqrt{x_n + h^2}\right)^2$$

for a convergent process.

4.5.* Write down the linear difference operator $\mathscr{L}_h$ associated with the backward Euler method and, working from first principles, find the precise form of the leading term in the LTE, i.e. find its order and error constant.

4.6.* Investigate the consistency of the LMM

$$x_{n+2} - x_n = \tfrac{1}{4}h(3f_{n+1} - f_n).$$

4.7.* What values should the parameters a and b have so that the following LMMs are consistent:

(a) $x_{n+2} - ax_{n+1} - 2x_n = hbf_n,$

(b) $x_{n+2} + x_{n+1} + ax_n = h(f_{n+2} + bf_n).$

4.8.* If β_1 were to be kept free in Example 4.10, say $\beta_1 = \theta$, show that this would lead to the one-parameter family of order 1 methods:

$$x_{n+2} - x_n = h(\theta f_{n+1} + (2 - \theta)f_n).$$

What is the error constant of this method? What value of θ produces the highest order?

4.9.** Show that the LMM $x_{n+1} = x_n + 2hf_n$ is not consistent and that, when used to solve the IVP $x'(t) = 1$ for $t \in (0, 1]$, $x(0) = 0$, it leads to $x_n = 2nh$. Hence, by writing down the GE $x(t_n) - x_n$ when $t_n = 1$, show that the method is not convergent.

4.10.* Determine the order and error constant of the LMM

$$3x_{n+2} - 4x_{n+1} + x_n = 2hf_{n+2}, \qquad (4.25)$$

which is known as the "two-step backward differentiation formula": BDF(2).

4.11.* Determine the two-step LMM of the form

$$x_{n+2} + \alpha_1 x_{n+1} - a x_n = h\beta_2 f_{n+2}$$

that has maximum order when a is retained as a free parameter.

4.12.** Show that an LMM of the form

$$x_{n+2} - x_{n+1} = h(\beta_2 f_{n+2} + \beta_1 f_{n+1} + \beta_0 f_n)$$

has maximum order when $\beta_2 = 5/12$, $\beta_1 = 2/3$, and $\beta_0 = -1/12$. What is the error constant of this method (known as the two-step Adams–Moulton method (AM(2))?

4.13.*** Write down the linear difference operator $\mathscr{L}_h$ associated with Simpson's rule

$$x_{n+2} - x_n = \tfrac{1}{3}h(f_{n+2} + 4f_{n+1} + f_n).$$

Use the Taylor expansions of $z(t+2h)$ and $z(t)$ about the point $t+h$:

$$z(t + 2h) = z(t + h) + hz'(t + h) + \tfrac{1}{2!}h^2 z''(t + h) + \ldots,$$
$$z(t) = z(t + h) - hz'(t + h) + \tfrac{1}{2!}h^2 z''(t + h) - \ldots,$$

with similar expansions for $z'(t + 2h)$ and $z'(t)$, to show that it has order $p = 4$ (the maximum possible for a two-step method) with error constant $C_5 = -1/90$.

What is the advantage of using these non-standard expansions?

4.14.*** Show that the general coefficient C_m $(m > 0)$ in the expansion (4.17) is given by

$$C_m = \sum_{j=0}^{2} \left[\frac{1}{m!} j^m \alpha_j - \frac{1}{(m-1)!} j^{m-1} \beta_j \right], \qquad (4.26)$$

where $\alpha_2 = 1$ is the normalizing condition.

Use this expression to verify the error constants found in Examples 4.4 and 4.10.

4.15.** Show that the conditions (4.15) for consistency are equivalent to the equations $\mathscr{L}_h z(t) = 0$ with the choices $z(t) = 1$ and $z(t) = t$.

Hence prove that $\mathscr{L}_h z(t) = 0$ whenever $z(t)$ is a linear function of t.

4.16.** If $x'(t) = g(t)$, $t \in [a, b]$, with $x(a) = 0$, show that $x(b) = \int_a^b g(t)\,dt$.

Use the trapezoidal rule with $h = 1/2$ to compute an approximate value for the integral $\int_0^1 t^3\,dt$ and compare with the exact value.

4.17.*** Complete the details of the proof of the second part of Theorem 4.8. Explain, in particular, why it is necessary to invoke l'Hôpital's rule.

4.18.** Determine the quadratic polynomial[5] $p(t)$ that satisfies the conditions

$$p(t_{n+1}) = x_{n+1}, \quad p'(t_{n+1}) = f_{n+1}, \quad p'(t_n) = f_n.$$

Show that $x_{n+2} = p(t_{n+2})$ leads to the AB(2) method (4.10).

4.19.** Find a quadratic polynomial $p(t)$ that satisfies the conditions

$$p(t_n) = x_n, \quad p'(t_n) = f_n, \quad p'(t_{n+1}) = f_{n+1}.$$

Show that $x_{n+1} = p(t_{n+1})$ leads to the trapezoidal rule (4.7).

---

[5]The construction of $p(t)$ can be simplified by seeking coefficients A, B, and C such that $p(t) = A + B(t - t_{n+1}) + C(t - t_{n+1})^2$.

5

Linear Multistep Methods—II: Convergence and Zero-Stability

5.1 Convergence and Zero-Stability

Means of determining the coefficients in LMMs were described in Chapter 4 and criteria now need to be established to identify those methods that are practically useful. In this section we describe some of the behaviour that should be expected of methods (in general) and, in subsequent sections, indicate how this behaviour can be designed into LMMs.

A basic requirement of any method is that its solutions should *converge* to those of the corresponding IVP as $h \to 0$. To formalize this concept, our original specification in Definition 2.3 must be tailored to accommodate the additional starting values needed for k-step LMMs.

In order to solve the IVP

$$\left. \begin{array}{l} x'(t) = f(t, x(t)), \quad t > t_0 \\ x(t_0) = \eta \end{array} \right\} \tag{5.1}$$

over some time interval $t \in [t_0, t_f]$, we choose a step-size h, a k-step LMM

$$x_{n+k} + \alpha_{k-1} x_{n+k-1} + \cdots + \alpha_0 x_n = h\big(\beta_k f_{n+k} + \beta_{k-1} f_{n+k-1} + \cdots + \beta_0 f_n\big), \tag{5.2}$$

and starting values

$$x_0 = \eta_0, \quad x_1 = \eta_1, \ldots, \quad x_{k-1} = \eta_{k-1}. \tag{5.3}$$

D.F. Griffiths, D.J. Higham, *Numerical Methods for Ordinary Differential Equations*, Springer Undergraduate Mathematics Series, DOI 10.1007/978-0-85729-148-6_5, © Springer-Verlag London Limited 2010

The values $x_k, x_{k+1}, ..., x_N$ are then calculated from (5.2) with $n = 0 : N - k$, where $Nh = t_f - t_0$.

Our aim is to develop methods with as wide a range of applicability as possible; that is, we expect the methods we develop to be capable of solving all IVPsof the form (5.1) that have unique solutions. We have no interest in methods that can solve only particular IVPsor particular types of IVP.

In addition to the issues related to convergence that we previously discussed in Section 2.4 (for Euler's method) and Section 3.4 (for TS(p) methods), the main point to be borne in mind when dealing with k-step LMMs concerns the additional starting values (5.3). These will generally contain some level of error[1] which must tend to zero as $h \to 0$, so that

$$\lim_{h \to 0} \eta_j = \eta, \quad j = 0 : k - 1. \tag{5.4}$$

In practice, the additional starting values $x_1, \ldots, x_{k-1}$ would be calculated using an appropriate numerical method. For example, condition (5.4) would be satisfied if we used $k - 1$ steps of Euler's method.

Definition 5.1 (Convergence)

The LMM (5.2) with starting values satisfying (5.4) is said to be *convergent* if, for all IVPs (5.1) that possess a unique solution $x(t)$ for $t \in [t_0, t_f]$,

$$\lim_{\substack{h \to 0 \\ nh = t^* - t_0}} x_n = x(t^*) \tag{5.5}$$

holds for all $t^* \in [t_0, t_f]$.

The next objective is to establish conditions on the coefficients of the general LMM that will ensure convergence. Theorem 4.8 shows that consistency is a necessary prerequisite, but Example 4.11 strongly suggests that, on its own, it is not sufficient. Recall that consistency implies that $\rho(1) = 0$ and $\rho'(1) = \sigma(1)$:

$$\sum_{j=0}^{k} \alpha_j = 0, \quad \sum_{j=0}^{k} j\alpha_j = \sum_{j=0}^{k} \beta_j, \tag{5.6}$$

with our normalizing condition $\alpha_k = 1$.

Example 5.2

Explain the the non-convergence of the two-step LMM

$$x_{n+2} + 4x_{n+1} - 5x_n = h(4f_{n+1} + 2f_n)$$

---

[1]Although it is natural to choose $\eta_0 = \eta$, the given starting value for the ODE, this is not necessary for convergence.

that was observed in Example 4.11 by applying the method to the (trivial) IVP $x'(t) = 0$ with initial condition $x(0) = 1$. Use starting values $x_0 = 1$ and $x_1 = 1 + h$ (recall that the definition of convergence requires only that $x_1 \to x(0)$ as $h \to 0$).

The method becomes, in this case,

$$x_{n+2} + 4x_{n+1} - 5x_n = 0. \tag{5.7}$$

This is a two-step constant-coefficient difference equation whose auxiliary equation is

$$r^2 + 4r - 5 = (r - 1)(r + 5).$$

The general solution of the difference equation (see Appendix D) is

$$x_n = A + B(-5)^n,$$

where A and B are arbitrary constants. The initial condition $x_0 = 1$ implies that $A + B = 1$ and $x_1 = 1 + h$ leads to

$$1 + h = A + B(-5).$$

These solve to give $B = -h/6$ and $A = 1 - h/6$, and so

$$x_n = 1 + \tfrac{1}{6}h[1 - (-5)^n].$$

It is the presence of the term $(-5)^n$ that causes the disaster: suppose, for instance, that $t = 1$, so $nh = 1$, then

$$h|(-5)^n| = \frac{1}{n}5^n \to \infty \quad \text{as } h \to 0.$$

When $h = 0.1$, for example, $n = 10$ and $\frac{1}{n}5^n \approx 10^6$, so the divergence of x_n is, potentially, very rapid. $\qquad\square$

The auxiliary equation of (5.7) is the first characteristic polynomial, $\rho(r)$, of the LMM. It has the property $\rho(1) = 0$, i.e. $r = 1$ is a root of $\rho(r)$ for all consistent methods. This means that the first characteristic polynomial of any consistent two-step LMM will factorize as

$$\rho(r) = (r - 1)(r - a)$$

for some value of a (in the previous example we had $a = -5$, which led to the trouble). A method with this characteristic polynomial applied to $x'(t) = 0$ would give a general solution

$$x_n = A + Ba^n,$$

which suggests that we should restrict ourselves to methods for which $|a| \leq 1$ so that $|a|^n$ does not go to infinity with n. This turns out to be not quite sufficient, as the next example shows.

Example 5.3

Investigate the convergence of the three-step LMM

$$x_{n+3} + x_{n+2} - x_{n+1} - x_n = 4hf_n$$

when applied to the model problem $x'(t) = 0$, $x(0) = 1$ with starting values $x_0 = 1$, $x_1 = 1 - h$, and $x_2 = 1 - 2h$.

Establishing consistency of the method is left to Exercise 5.7.

The homogeneous difference equation $x_{n+3} + x_{n+2} - x_{n+1} - x_n = 0$ has auxiliary equation

$$\rho(r) = (r - 1)(r + 1)^2$$

and, therefore, its general solution is

$$x_n = A + (B + Cn)(-1)^n$$

(see Appendix D). With the given starting values, the solution can be shown to be

$$x_n = 1 - h + (-1)^n(h - t_n), \tag{5.8}$$

while the exact solution of the IVP is, of course, $x(t) = 1$. Thus, for example, at $t = t^* = 1$, $x(1) = 1$ while, $|x_n - 1 + h| = 1 - h$ *for all values of h* with $t^* = nh = 1$. The GE, therefore, has the property $|x(1) - x_n| \to 1$ and does not tend to zero as $h \to 0$. Hence, the method is consistent but not convergent. $\square$

This leads to the following definition.

Definition 5.4 (Root Condition)

A polynomial is said to satisfy the *root condition* if all its roots lie within or on the unit circle, with those on the boundary being simple. In other words, all roots satisfy $|r| \leq 1$ and any that satisfy $|r| = 1$ are simple.[2]

A polynomial satisfies the *strict root condition* if all its roots lie inside the unit circle; that is, $|r| < 1$.

Thus, $\rho(r) = r^2 - 1$ satisfies the root condition, while $\rho(r) = (r - 1)^2$ does not.

Definition 5.5 (Zero-Stability)

An LMM is said to be *zero-stable* if its first characteristic polynomial $\rho(r)$ satisfies the root condition.

---

[2] We say that λ is a simple root of $\rho(r)$ if $\lambda - r$ is a factor of $\rho(r)$, but $(\lambda - r)^2$ is not.

All consistent one-step LMMs have first characteristic polynomial $\rho(r) = r - 1$ and so satisfy the root condition automatically, which is why this notion was not needed in the study of Euler's method or the Taylor series methods. We are now in a position to state Dahlquist's celebrated theorem.

Theorem 5.6 (Dahlquist (1956))

An LMM is convergent if, and only if, it is both consistent and zero-stable.

The main purpose of this theorem is to filter out the many LMMs that fail its conditions; we are left to focus our attention on those that pass—the convergent methods. An explanation of why both conditions together are sufficient is given in Section 5.3 in the context of a scalar linear IVP.

Zero-stability places a significant restriction on the attainable order of LMMs, as the next theorem attests—recall that the order of an implicit (explicit) k-step LMM could be as high as $2k$ $(2k - 1)$.

Theorem 5.7 (First Dahlquist Barrier (1959))

The order p of a stable k-step LMM satisfies

1. $p \leq k + 2$ if k is even;

2. $p \leq k + 1$ if k is odd;

3. $p \leq k$ if $\beta_k \leq 0$ (in particular for all explicit methods).

Proofs of Theorems 5.6 and 5.7 were originally given in the landmark papers of Dahlquist [14, 15]. They may also be found in the book of Hairer et al. [28].

Had we been armed with this barrier theorem earlier, the method described in Example 4.11 could have been dismissed immediately, since it violates part 3 of the theorem (it is explicit with $k = 2$ and $p = 3$).

5.2 Classic Families of LMMs

The first four families of classical methods we describe below are constructed by choosing methods that have a particularly simple first characteristic polynomial that will automatically be zero-stable. The remaining coefficients are chosen so that the resulting method will have maximum order.

1. *Adams–Bashforth (1883):* These have first characteristic polynomials

$$\rho(r) = r^k - r^{k-1},$$

which have a simple root $r = 1$ and a root of multiplicity $(k-1)$ at $r = 0$. Methods in this family are also explicit, so

$$x_{n+k} - x_{n+k-1} = h\big(\beta_{k-1}f_{n+k-1} + \cdots + \beta_0 f_n\big)$$

and the k coefficients $\beta_0, \ldots, \beta_{k-1}$ are chosen so that $C_0 = C_1 = \cdots = C_{k-1} = 0$ and give, therefore, a method of order $p = k$. This is the highest possible order of a zero-stable explicit k-step method (part 3 of Theorem 5.7) and is attained despite imposing severe restrictions on its structure.

We have already encountered the $k = 1$ member—Euler's method—and the AB(2) method (4.10) for $k = 2$. With $k = 3$ the AB(3) method is

$$x_{n+3} - x_{n+2} = \tfrac{1}{12}h\big(23f_{n+2} - 48f_{n+1} + 5f_n\big), \tag{5.9}$$

which has order $p = 3$ and error constant $C_3 = \tfrac{3}{8}$.

2. *Adams–Moulton (1926):* These are implicit versions of the Adams–Bashforth family, having the form

$$x_{n+k} - x_{n+k-1} = h\big(\beta_k f_{n+k} + \cdots + \beta_0 f_n\big).$$

There is one more coefficient than the corresponding Adams–Bashforth method so we can attain order $k + 1$. For $k = 1$ we have the trapezoidal rule (4.7). The two-step Adams–Moulton method (AM(2)—see Exercise 4.12) has order 3 and not the maximum possible order (4) for an implicit two-step method implied by Theorem 5.7. For $k = 3$, the AM(3) method

$$x_{n+3} - x_{n+2} = \tfrac{1}{24}h\big(9f_{n+3} + 19f_{n+2} - 5f_{n+1} + f_n\big)$$

has order 4 and error constant $C_4 = -19/720$.

3. *Nyström method (1925):s* These are explicit methods with $k \geq 2$ having first characteristic polynomial $\rho(r) = r^k - r^{k-2}$, so take the general form

$$x_{n+k} - x_{n+k-2} = h\big(\beta_{k-1}f_{n+k-1} + \cdots + \beta_0 f_n\big).$$

They have the same number of free coefficients as the AB family and are chosen to achieve order $p = k$, the same order as the corresponding AB method. The two-step version is the second-order mid-point rule (4.21).

4. *Milne–Simpson (1926):* These are the implicit analogues of Nyström methods:

$$x_{n+k} - x_{n+k-2} = h\big(\beta_k f_{n+k} + \cdots + \beta_0 f_n\big).$$

The simplest member of the family is Simpson's rule (see Exercise 4.13), which has $k = 2$ and order $p = 4$. The high degree of symmetry present in the structure of the method allows it to attain the maximum possible order for a zero-stable two-step method (unlike AM(2) above).

5. *Backward differentiation formulas (BDFs—1952).* These form a generalization of the backward Euler method (see Table 4.2). The simplest possible form is chosen for the second characteristic polynomial consistent with the method being implicit: $\sigma(r) = \beta_k r^k$. Thus,

$$x_{n+k} + \alpha_{k-1} x_{n+k-1} + \cdots + \alpha_0 x_n = h\beta_k f_{n+k}$$

and the $(k+1)$ free coefficients are chosen to achieve order k—not the higher order $(k+2)$ that can be achieved with the most general implicit zero-stable k-step LMMs. Despite this, they form an important family because they have compensating strengths, as we shall see in Chapter 6.

It may be shown (see Iserles [39, Lemma 2.3], for instance) that the first characteristic polynomial, $\rho(r)$, of these methods is given by

$$\rho(r) = \frac{1}{c} \sum_{j=1}^{k} \frac{1}{j} r^{k-j} (r-1)^j, \qquad c = \sum_{j=1}^{k} \frac{1}{j}. \tag{5.10}$$

Zero-stability is easily checked for each k—the BDF(2) method of Exercise 4.10 is stable, as are other members up to $k = 6$; thereafter they are all unstable.

5.3 Analysis of Errors: From Local to Global

The convergence of Euler's method was analysed in Theorem 2.4 for the special case when it was applied to the IVP

$$\begin{aligned} x'(t) &= \lambda x(t) + g(t), \quad 0 < t \le t_f \\ x(0) &= 1. \end{aligned} \tag{5.11}$$

In this section we shall indicate the additional features that have to be accounted for when analysing the convergence of k-step LMMs with $k > 1$. To get across the key concepts, it is sufficient to look at the behaviour of two-step methods.

As discussed in Section 4.3, LMMs are constructed from Taylor expansions in such a manner that their associated linear difference operators have the property

$$\mathscr{L}_h z(t) = C_{p+1} h^{p+1} z^{(p+1)}(t) + \cdots ,$$

where z is any $(p+1)$-times continuously differentiable function.

The LTE, denoted by T_{n+2}, is defined to be the value of the left-hand side of this expression when we replace z by x, the exact solution of our IVP. So, for our two-step method we define, at $t = t_{n+2}$,

$$T_{n+2} = \mathscr{L}_h x(t_n). \qquad (5.12)$$

When $x(t)$ is a $(p+1)$-times continuously differentiable function we have

$$T_{n+2} = C_{p+1} h^{p+1} x^{(p+1)}(t) + \cdots ,$$

so that $T_{n+2} = \mathcal{O}(h^{p+1})$. Let $y_n = x(t_n)$ (for all n) denote the exact solution of the IVP (5.11) at a typical grid point t_n. The general two-step LMM applied to the IVP (5.11) leads to

$$x_{n+2} + \alpha_1 x_{n+1} + \alpha_0 x_n =$$
$$h\lambda\big(\beta_2 x_{n+2} + \beta_1 x_{n+1} + \beta_0 x_n\big)$$
$$+ h(\beta_2 g(t_{n+2}) + \beta_1 g(t_{n+1}) + \beta_0 g(t_n)). \quad (5.13)$$

With Definition 4.2 for $\mathscr{L}_h$ and equation (5.12) we find that the exact solution satisfies the same equation with the addition of T_{n+2} on the right:

$$y_{n+2} + \alpha_1 y_{n+1} + \alpha_0 y_n =$$
$$h\lambda\big(\beta_2 y_{n+2} + \beta_1 y_{n+1} + \beta_0 y_n\big)$$
$$+ h(\beta_2 g(t_{n+2}) + \beta_1 g(t_{n+1}) + \beta_0 g(t_n)) + T_{n+2}. \quad (5.14)$$

Subtracting (5.13) from (5.14), the GE $e_n = x(t_n) - x_n \equiv y_n - x_n$ is found to satisfy the difference equation

$$(1 - h\lambda\beta_2)e_{n+2} + (\alpha_1 - h\lambda\beta_1)e_{n+1} + (\alpha_0 - h\lambda\beta_0)e_n = T_{n+2}, \qquad (5.15)$$

with the starting values $e_0 = 0$ and $e_1 = x(t_1) - \eta_1$, which may be assumed to be small (the definition of convergence stipulates that the error in starting values must tend to zero as $h \to 0$). This is the equation that governs the way "local" errors T_{n+2} accumulate into the "global" error $\{e_n\}$. Usage of the terms "local" and "global" reflects the fact that the LTE T_{n+2} can be calculated locally (for each time t_n) via Equation (5.12), whereas the GE results from the overall accumulation of all LTEs—a global process.

In order to simplify the next stage of the analysis, let us assume that the LTE terms T_{n+2} are constant: $T_{n+2} = T$, for all n. As discussed in Appendix D, the general solution of (5.15) is then comprised of two components:

1. A particular solution, $e_n = P$, which is constant. Substituting $e_{n+2} = e_{n+1} = e_n = P$ into (5.15) and using the consistency condition $1 + \alpha_1 + \alpha_0 = 0$, we find that[3]

$$P = \frac{T}{h\lambda\sigma(1)};$$

so, if $T = \mathcal{O}(h^{p+1})$, it follows that $P = \mathcal{O}(h^p)$.

2. The general solution of the homogeneous equation. If the auxiliary equation

$$(1 - h\lambda\beta_2)r^2 + (\alpha_1 - h\lambda\beta_1)r + (\alpha_0 - h\lambda\beta_0) = 0$$

has distinct roots, $r_1 \neq r_2$, then this contribution is

$$Ar_1^n + Br_2^n,$$

where A and B are arbitrary constants.

Thus, the general solution for the GE is, in this case,

$$e_n = Ar_1^n + Br_2^n + P. \tag{5.16}$$

The constants A and B are determined from the starting values. As $h \to 0$, the roots r_1 and r_2 tend to the roots of the first characteristic polynomial $\rho(r)$. If the root condition in Definition 5.4 is violated, then their contribution will completely swamp the GE and lead to divergence. The LTE contributes to the GE through the term $P = \mathcal{O}(h^p)$, so consistency ($p > 0$) ensures that this tends to zero with h.

Consistency of an LMM ensures that the local errors are small, while zero-stability ensures that they propagate so as to give small GEs.

5.4 Interpreting the Truncation Error

We extend the discussion in the previous subsection so as to give an alternative—and perhaps more intuitive—interpretation of the LTE (5.12) of an LMM. To simplify the presentation we will again restrict ourselves to the case of a linear ODE solved with a two-step method, but the conclusions are valid more generally.

---

[3]The presence of the factor $\sigma(1)$ in the denominator induces many writers to define the error constant of an LMM to be $C_{p+1}/\sigma(1)$ and not C_{p+1} as we have done. We shall refer to $C_{p+1}/\sigma(1)$ as the scaled error constant; it is seen on page 89 to be a key feature of Dahlquist's Second Barrier Theorem.

Suppose we were to compute x_{n+2} from (5.13) when the past values were exact—that is, $x_{n+1} = y_{n+1}$ and $x_n = y_n$ (recall that $y_{n+j} = x(t_{n+j})$)—and we denote the result by $\widetilde{x}$:

$$\widetilde{x}_{n+2} + \alpha_1 x_{n+1} + \alpha_0 x_n =$$
$$h\lambda\big(\beta_2 \widetilde{x}_{n+2} + \beta_1 x_{n+1} + \beta_0 x_n\big)$$
$$+ h\big(\beta_2 g(t_{n+2}) + \beta_1 g(t_{n+1}) + \beta_0 g(t_n)\big).$$

Subtracting this from (5.14) gives

$$(1 - h\lambda\beta_2)(y_{n+2} - \widetilde{x}_{n+2}) = T_{n+2}.$$

1. In the explicit case ($\beta_2 = 0$), $T_{n+2} = y_{n+2} - \widetilde{x}_{n+2}$: the LTE is the error committed in one step on the assumption that the back values are exact.

2. In the implicit case ($\beta_2 \neq 0$), the binomial expansion:

$$(1 - h\lambda\beta_2)^{-1} = 1 + h\lambda\beta_2 + \mathcal{O}(h^2) = 1 + \mathcal{O}(h)$$

and the fact that $T_{n+2} = \mathcal{O}(h^{p+1})$ for a method of order p shows that $(1 + \mathcal{O}(h))T_{n+2} = y_{n+2} - \widetilde{x}_{n+2}$ and hence

$$T_{n+2} = y_{n+2} - \widetilde{x}_{n+2} + \mathcal{O}(h^{p+2}).$$

So, the *leading term* in the LTE is the error committed in one step on the assumption that the back values are exact.

The assumption that back values are exact is known as the *localizing assumption* (Lambert [44, p. 56]).

EXERCISES

5.1.* Investigate the zero-stability of the two-step LMMs

(a) $x_{n+2} - 4x_{n+1} + 3x_n = -2hf_n$,

(b) $3x_{n+2} - 4x_{n+1} + x_n = ahf_n$.

Are there values of a in part (b) for which the method is convergent?

5.2.** Show that the order of the LMM

$$x_{n+2} + (b-1)x_{n+1} - bx_n = \tfrac{1}{4}h\left[(b+3)f_{n+2} + (3b+1)f_n\right]$$

is 2 if $b \neq -1$ and 3 if $b = -1$.

Show that the method is not zero-stable when $b = -1$, and illustrate the divergence by applying it to the IVP $x'(t) = 0$, $x(0) = 0$ with starting values $x_0 = 0$ and $x_1 = h$.[4]

5.3.** Show that the LMM

$$x_{n+2} + 2ax_{n+1} - (2a + 1)x_n = h((a + 2)f_{n+1} + af_n)$$

has order 2 in general and express the error constant in terms of a.

Deduce that there is one choice of the parameter a for which the method has order 3 but that this method is not zero-stable. How is this method related to Example 4.11?

What value of a leads to a convergent method with smallest possible error constant?

5.4.** Consider the one-parameter family of LMMs

$$x_{n+2} + 4ax_{n+1} - (1 + 4a)x_n = h((1 + a)f_{n+1} + (1 + 3a)f_n).$$

for solving the ODE $x'(t) = f(t, x(t))$.

(a) Determine the error constant for this family of methods and identify the method of highest order.

(b) For what values of a are members of this family convergent?

Is the method of highest order convergent? Explain your answer carefully.

5.5.* Prove that any consistent one-step LMM is always zero-stable.

5.6.* Suppose that the first characteristic polynomial $\rho(r)$ of a convergent LMM is a quadratic polynomial. Prove that the most general form of ρ is

$$\rho(r) = (r - 1)(r - a)$$

for a parameter $a \in [-1, 1)$.

5.7.** For Example 5.3, verify that

(a) the method is consistent;

(b) the numerical solution at time t_n is given by the expression (5.8);

(c) the method is not zero-stable.

---

[4]This example illustrates the conflicting requirements of trying to achieve maximum order of consistency while maintaining zero-stability. When $b = -1$ the characteristic polynomials ρ and σ have a common factor—such methods are called reducible LMMs.

5.8.* Prove that the method

$$x_{n+3} + x_{n+2} - x_{n+1} - x_n = h(f_{n+3} + f_{n+2} + f_{n+1} + f_n)$$

is consistent but not convergent.

5.9.* Investigate the convergence, or otherwise, of the method

$$x_{n+3} - x_{n+2} + x_{n+1} - x_n = \tfrac{1}{2}h(f_{n+3} + f_{n+2} + f_{n+1} + f_n).$$

5.10.* Prove that $\sigma(1) \neq 0$ for a convergent k-step LMM.

5.11.* Consider the two-parameter family of LMMs

$$x_{n+2} - x_n = h\left(\beta_1 f_{n+1} + \beta_0 f_n\right).$$

Show that there is a one-parameter family of convergent methods of this type.

Determine the error constant for each of the convergent methods and identify the method of highest order. What is its error constant?

5.12.*** Apply the LMM $x_{n+2} - 2x_{n+1} + x_n = h(f_{n+1} - f_n)$ to the IVP $x'(t) = -x(t)$ with $x(0) = 1$ and the starting values $x_0 = x_1 = 1$ (note that these satisfy our criterion (5.4)). Show that the exact solution of the LMM in this case is $x_n = A + B(1 - h)^n$. Find the values of A and B and discuss the convergence of the method by directly comparing the expressions for x_n and $x(t_n)$.

Extend your solution to the initial values $x_0 = 1$, $x_1 = 1 + ah$.

Relate your conclusions to the zero-stability of the method.

5.13.* Consider the family of LMMs

$$x_{n+2} + (\theta - 2)x_{n+1} + (1 - \theta)x_n = \tfrac{1}{4}h\left((6 + \theta)f_{n+2} + 3(\theta - 2)f_n\right),$$

in which θ is a parameter.

Determine the order and error constant for the method and show that both are independent of θ.

For what range of θ values is the method convergent? Explain your answer.

5.14.** Determine the coefficients in the three-step BDF(3) method (see page 67) and calculate its error constant. Verify that it has order $p = 3$ and that the coefficients give rise to the first characteristic polynomial (5.10) with $k = 3$. Examine the zero-stability of the resulting method.

5.15.** The first characteristic polynomial of the general BDF(k) is given by (5.10). Use consistency of the method to derive an expression for the second characteristic polynomial $\sigma(r)$.

Verify that, when $k = 2$, these characteristic polynomials give rise to the BDF(2) method given in Exercise 4.10.

5.16.*** Use the starting values $e_0 = 0$ and $e_1 = x(h) - \eta_1$ to determine the constants A and B in (5.16) and hence show that

$$e_n = (x(h) - \eta_1)\frac{r_1^n - r_2^n}{r_1 - r_2} + P\left(1 - \frac{1 - r_2}{r_1 - r_2}r_1^n + \frac{1 - r_1}{r_1 - r_2}r_2^n\right).$$

If the first characteristic polynomial has roots $r = 1$ and $r = a$, then it may be shown that $r_1 = 1 + \mathcal{O}(h)$ and $r_2 = a + \mathcal{O}(h)$, where $-1 \leq a < 1$ and, consequently, $r_1 - r_2 = 1 - a + \mathcal{O}(h)$ is bounded away from zero as $h \to 0$. It then follows that $e_n = \mathcal{O}(h^p)$ if $P = \mathcal{O}(h^p)$ and the error in the additional starting value is of the same order: $\eta_1 = x(h) + \mathcal{O}(h^p)$.

6
Linear Multistep Methods—III:
Absolute Stability

6.1 Absolute Stability—Motivation

The study of convergence for LMMs involves the limit

$x_n \to x(t^*)$ when $n \to \infty$ and $h \to 0$ in such a way that $t_n = t_0 + nh = t^*$, where t^* is *fixed* value of the "time" variable t.

Thus, convergent methods generate numerical solutions that are arbitrarily close to the exact solution of the IVP provided that h is taken to be sufficiently small. Since non-convergent methods are of little practical use we shall henceforth assume that all LMMs used are convergent—they are consistent and zero-stable.

Being in possession of a convergent method may not be much comfort in practice if one ascertains that h needs to be particularly small to obtain results of even modest accuracy. If $h < 10^{-6}$, for instance, then more than 1 million steps would be needed in order to integrate over each unit of time; this may be entirely impractical. In this chapter we begin the investigation into the behaviour of solutions of LMMs when h is not arbitrarily small. We give a numerical example before pursing this question.

D.F. Griffiths, D.J. Higham, *Numerical Methods for Ordinary Differential Equations*,
Springer Undergraduate Mathematics Series, DOI 10.1007/978-0-85729-148-6_6,
© Springer-Verlag London Limited 2010

Example 6.1

Use Euler's method to solve the IVP (see Example 1.9)

$$x'(t) = -8x(t) - 40(3e^{-t/8} - 1), \qquad x(0) = 100.$$

We know that the exact solution of this problem is given by

$$x(t) = \frac{1675}{21}e^{-8t} + \frac{320}{21}e^{-t/8} + 5.$$

This function is shown in Figure 6.1. The temperature $x(t)$ drops quite rapidly from $100°C$ to room temperature (about $20°C$) and then falls slowly to the exterior temperature (about $5°C$).

Turning now to the numerical solution, Euler's method applied to this ODE gives

$$x_{n+1} = (1 - 8h)x_n + h\left(120e^{-t_n/8} + 40\right), \qquad n = 0, 1, 2, \ldots,$$

with $t_n = nh$ and $x_0 = 100$.

The leftmost graph in Figure 6.2 shows x_n versus t_n when $h = 1/3$. The numerical solution has ever-increasing oscillations—a classic symptom of numerical instability—reaching a maximum amplitude of 10^6. The numerical solution bears no resemblance to the exact solution shown in Figure 6.1.

Reducing h (slightly) to $1/5$ has a dramatic effect on the solution (middle graph): it now decays in time while continuing to oscillate until about $t \approx 2$, after which point it becomes a smooth curve.

When h is further reduced to $1/9$ the solution resembles the exact solution, but the solid and broken curves do not become close until $t \approx 0.5$.

In Figure 6.3 we plot the GEs (GE) $|x(nh) - x_n|$ on a log-linear scale.[1] The linear growth of the GE on this scale for $h = 1/3$ suggests exponential growth. In contrast, for $h = 1/5$, the error decays exponentially over the interval

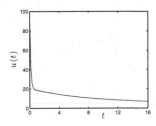

Fig. 6.1

The exact solution to Example 6.1

---

[1] If the GE were to vary exponentially with n, $e_n = cr^n$, then $\log e_n = n \log r + \log c$ and a graph of $\log e_n$ versus n would be a straight line with slope $\log r$. When $r > 1$ the slope would be positive, corresponding to exponential growth, while the slope would be negative if $0 < r < 1$, corresponding to exponential decay.

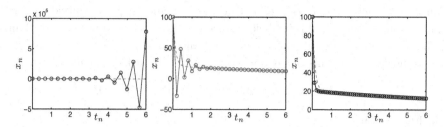

Fig. 6.2 Solutions to Example 6.1 using Euler's method with $h = 1/3$ (left), $h = 1/5$ (middle) and $h = 1/9$ (right). The exact solution of the IVP is shown as a broken curve

$0 \leq t \leq 4$ where it reaches a level of about about 10^{-3}. Were we to be interested in an accuracy of about $1°C$, this would be achieved at around $t = 1.8$ h.

For $h = 1/9$ there is a more rapid initial exponential decay of the error until it, too, levels out, at a similar value of about 10^{-3} at $t = 1.3$ h.

The theory that we shall present below can be used to show (see Example 6.7) that Euler's method suffers a form of instability for this IVP when $h \geq 1/4$; this is clearly supported by the numerical evidence we have presented. This is typical: Euler's method is of no practical use for problems with exponentially decaying solutions unless h is small enough. Furthermore, when h is chosen small enough to avoid exponential growth of the GE, the accuracy of the long-term solution may be much higher than is required.

This behaviour is caused by the exponent e^{-8t} in the complementary function; had it been e^{-80t} we would have needed to take h to be 10 times smaller and we would have generated, as a consequence, a numerical solution that was 10 times more accurate. Thus, in this problem, we are forced to choose a small value of h in order to avoid instability and, while this will produce a solu-

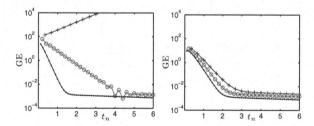

Fig. 6.3 The GEs associated with the solutions to Example 6.1 using Euler's method (left) and the backward Euler method (right) with $h = 1/3$ (+), $h = 1/5$ (○), and $h = 1/9$ (●)

tion of high accuracy, it is likely to be more accurate than required by the application—the method is inefficient since we have expended more effort than necessary when only moderate accuracy is called for.

We now contrast this behaviour with that of the backward Euler method:

$$x_{n+1} = x_n - 8hx_{n+1} + h\big(120e^{-t_{n+1}/8} + 40\big), \qquad n = 0, 1, 2, \ldots,$$

which is rearranged to read

$$x_{n+1} = \frac{1}{1 + 8h}\big[x_n + h(120e^{-t_{n+1}/8} + 40)\big], \qquad n = 0, 1, 2, \ldots,$$

with $t_{n+1} = (n+1)h$ and $x_0 = 100$.

Solutions corresponding to $h = 1/3, 1/5, 1/9$ are shown in Figure 6.4. Comparing with Figure 6.2 it is evident that there are no oscillations and the numerical solutions are reasonable approximations of the exact solution for each of the values of h. The behaviour is such that small changes to h lead to small changes in the numerical solutions (and associated GEs)—this is a desirable, stable feature.

The *local* truncation errors for forward and backward Euler methods are equal but opposite in sign. However, the way that these errors propagate is clearly different. In this example the backward Euler method is much superior to that of the Euler method since we are able to choose h on grounds of accuracy alone without having to be concerned with exponentially growing oscillations for larger values of h.   ☐

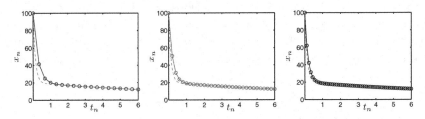

Fig. 6.4 Solutions to Example 6.1 using the backward Euler method with $h = 1/3$ (left), $h = 1/5$ (middle), and $h = 1/9$ (right). The exact solution of the IVP is shown as a broken curve

Example 6.2

Use the forward and backward Euler methods to solve the IVP (see (1.16))

$$x'(t) = -\tfrac{1}{8}(x(t) - 5 - 5025e^{-8t}), \qquad x(0) = 100.$$

Compare the results produced with those of the previous example which has the same exact solution (see Example 1.9).

The forward and backward Euler methods are, respectively,

$$x_{n+1} = (1 - \tfrac{1}{8}h)x_n - h(5 - 5025e^{-8t_n}),$$

$$x_{n+1} = \frac{1}{1 + h/8}\left[x_n - h(5 - 5025e^{-8t_{n+1}})\right],$$

with $x_0 = 100$ in each case. The GEs obtained when these methods are deployed with $h = 1/3$ and $1/9$ are shown in Figure 6.5. In both cases the GEs are quite large over the interval of integration and decay slowly (proportional to $e^{-t/8}$). However, the most important feature in the present context is that there is no indication of the exponential growth that we saw in the previous example with Euler's method with $h = 1/3$. In this example the forcing function decays rapidly relative to the solutions of the homogeneous equation (e^{-8t} *versus* $e^{-t/8}$), whereas the roles are reversed in Example 6.1. This suggests that the exponentially growing oscillations observed in Figure 6.2 with $h = 1/3$ are associated with rapidly decaying solutions of the homogeneous equation. This is the motivation for studying homogeneous equations in the remainder of this chapter.   □

6.2 Absolute Stability

The numerical results for Examples 6.1 and 6.2 lead us to *absolute stability theory*, in which we examine the effect of applying (convergent) LMMs to the *model scalar problem*

$$x'(t) = \lambda x(t), \tag{6.1}$$

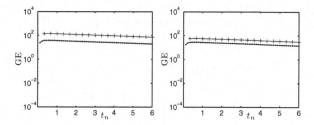

Fig. 6.5 The GEs associated with the solutions to Example 6.2 using Euler's method (left) and the backward Euler method (right) with $h = 1/3$ (+) and $h = 1/9$ (•)

in which λ may be complex[2] and has *negative real part*: $\Re(\lambda) < 0$. The general solution has the form $x(t) = c\,\mathrm{e}^{\lambda t}$ in which c is an arbitrary constant. Hence, $x(t) \to 0$ as $t \to \infty$, regardless of the value of c.

Our aim is to determine those LMMs which, when applied to (6.1), give solutions $\{x_n\}$ that also tend to zero as $t_n \to \infty$ with a given fixed step size h. Note that this property is different from that used by convergence theory, which also required the limit $n \to \infty$, since h is now fixed. This notion has proved, perhaps surprisingly, to be both important and useful over many years and we formalize our aspirations by the following definition.

Definition 6.3 (Absolute Stability)

An LMM is said to be *absolutely stable* if, when applied to the test problem $x'(t) = \lambda x(t)$ with $\Re(\lambda) < 0$ and a given value of $\widehat{h} = h\lambda$,[3] its solutions tend to zero as $n \to \infty$ for any choice of starting values.

Our definition of absolute stability is motivated by the idea of asking for the numerical method to reproduce the long-term behaviour of the model ODE (6.1). However, from Section 5.3 it should be clear that this condition is very similar to the requirement that the GE should be damped as time increases— this is the property that we looked at in Example 6.1. Hence, absolute stability is an important factor in the control of the GE. Applying the general two-step LMM (see Equation (4.13)) to the ODE $x'(t) = \lambda x(t)$ we have

$$x_{n+2} + \alpha_1 x_{n+1} + \alpha_0 x_n = h\lambda(\beta_2 x_{n+2} + \beta_1 x_{n+1} + \beta_0 x_n),$$

which can be rearranged to give the two-step linear difference equation

$$(1 - \widehat{h}\beta_2)x_{n+2} + (\alpha_1 - \widehat{h}\beta_1)x_{n+1} + (\alpha_0 - \widehat{h}\beta_0)x_n = 0. \tag{6.2}$$

This equation is relatively easy to analyse since it is a homogeneous linear difference equation with constant coefficients (see Appendix D). It has solutions of the form $x_n = ar^n$, where r is a root of the auxiliary equation

$$(1 - \widehat{h}\beta_2)r^2 + (\alpha_1 - \widehat{h}\beta_1)r + (\alpha_0 - \widehat{h}\beta_0) = 0. \tag{6.3}$$

We denote the polynomial on the left-hand side by $p(r)$. Notice that

$$p(r) = \rho(r) - \widehat{h}\sigma(r).$$

---

[2]The reason for allowing λ to be complex will become clear in Chapter 7 during the application to systems of differential equations.

[3]We introduce the single parameter $\widehat{h}$ since the parameters h and λ occur only as the product $h\lambda$.

We refer to $p(r)$ as the *stability polynomial* of the LMM; it is a polynomial of degree 2 whose coefficients depend (linearly) on the parameter $\widehat{h}$.

This stability polynomial will have two roots, r_1 and r_2, and so (6.2) will have the general solution

$$x_n = ar_1^n + br_2^n,$$

for arbitrary constants a and b provided that $r_1 \neq r_2$.[4] In order to have $|x_n| \to 0$ as $n \to \infty$ for any choices of a and b, it is necessary to have $|r_1| < 1$ and $|r_2| < 1$, i.e., the polynomial $p(r)$ must satisfy the strict root condition (Definition 5.4). This gives the following result.

Lemma 6.4

An LMM is absolutely stable for a given value of $\widehat{h} = \lambda h$ if, and only if, its stability polynomial $p(r)$ satisfies the strict root condition (Definition 5.4).

An LMM will not, in general, be absolutely stable for every choice of $\widehat{h}$, so we are led to define the following.

Definition 6.5 (Region of Absolute Stability)

The set of values $\mathcal{R}$ in the complex $\widehat{h}$-plane for which an LMM is absolutely stable forms its *region of absolute stability*.

We must, therefore, address the question of whether, and for what values of $\widehat{h}$, the roots of $p(r)$ satisfy $|r| < 1$. Cases where λ is real (and negative) are easier to analyse so, for these, we define the following.

Definition 6.6 (Interval of Absolute Stability)

The *interval of absolute stability* of an LMM is the largest interval of the form $\mathcal{R}_0 = (\widehat{h}_0, 0)$, with $\widehat{h}_0 < 0$, for which the LMM is absolutely stable for all *real* values of $\widehat{h} \in \mathcal{R}_0$.

The interval of absolute stability is found by looking at the intersection of the region of absolute stability with the negative real $\widehat{h}$ axis.

---

[4]The case of equality is unimportant since it could be countered by making a small change to the value of h.

Example 6.7

Find the region of absolute stability of Euler's method: $x_{n+1} - x_n = hf_n$.

Applied to $x'(t) = \lambda x(t)$ we have $f_n \equiv f(t_n, x_n) = \lambda x_n$ so that

$$x_{n+1} = x_n + h\lambda x_n = (1 + \widehat{h})x_n.$$

This has stability polynomial $p(r) = r - 1 - \widehat{h}$ with the single root $r_1 = 1 + \widehat{h}$. The region of absolute stability is, therefore, the open disc $|1 + \widehat{h}| < 1$ whose boundary is the circle of radius 1 centred at $\widehat{h} = -1$. To see this, let $\widehat{h} = \widehat{x} + i\widehat{y}$, then the boundary equation $|1 + \widehat{h}|^2 = 1$ leads to $(\widehat{x} + 1)^2 + \widehat{y}^2 = 1$.

If $\widehat{h}$ is real, the interval of absolute stability is given by

$$-1 < 1 + \widehat{h} < 1,$$

which leads to $\widehat{h} \in (-2, 0)$. See Figure 6.6.   □

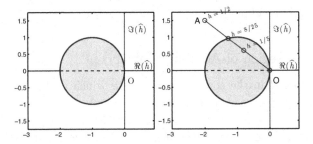

Fig. 6.6 The region of absolute stability for Euler's method (shaded) and the interval of absolute stability (broken line). On the right the line OA is the locus of the points $(-4h, 3h)$ for Example 6.8

If Euler's method were to be applied to $x'(t) = -8x(t)$, then $\widehat{h} = -8h$ and absolute stability would require $h < 1/4$. Similarly, for $x' = -80x$, we would have $\widehat{h} = -80h$ and absolute stability would require $h < 1/40$. This is the reason that sensible results were computed with Euler's method in Example 6.1 only when $h < 1/4$.

Example 6.8

What is the largest value of h that can be used so that Euler's method is absolutely stable when used to solve the ODE $x'(t) = \lambda x(t)$ with $\lambda = -3 + 4i$?

In this case $\widehat{h} = h(-4 + 3i)$, $|1 + \widehat{h}| = |1 + h(-4 + 3i)|$, and

$$|1 + \widehat{h}|^2 = |1 + h(-4 + 3i)|^2 = (1 - 4h)^2 + 9h^2.$$

Since $|1 + \widehat{h}| < 1$ is equivalent to $|1 + \widehat{h}|^2 - 1 < 0$ we calculate

$$|1 + \widehat{h}|^2 - 1 = h(-8 + 25h)$$

and conclude that the right-hand side will be negative if $h < 8/25$.

To interpret the situation geometrically, let $\widehat{h} = \widehat{x} + i\widehat{y}$, where $\widehat{x} = -4h$ and $\widehat{y} = 3h$. As h varies, the locus of the points $(-4h, 3h)$ is a straight line in the complex $\widehat{h}$-plane having equation $\widehat{y} = -\frac{4}{3}\widehat{x}$—this is shown as the line OA in Figure 6.6 (right). The indicated points on this line are at $h = 2/5, 8/25$ and $h = 1/2$. When $h = 8/25$ the point $(-4h, 3h)$ lies on the boundary of the region of stability.   □

Example 6.9

Find the region of absolute stability of the trapezoidal rule:

$$x_{n+1} - x_n = \tfrac{1}{2}h[f_{n+1} + f_n].$$

The stability polynomial is

$$p(r) = r - 1 - \tfrac{1}{2}\widehat{h}(r + 1),$$

which has the single root

$$r_1 = \frac{1 + \tfrac{1}{2}\widehat{h}}{1 - \tfrac{1}{2}\widehat{h}}.$$

It may be verified (Exercise 6.1) that $|r_1| < 1$ for all values of $\widehat{h}$ with negative real part, so the region of absolute stability is the entire left half plane and the interval of absolute stability is given by $\widehat{h} \in (-\infty, 0)$. See Figure 6.7.   □

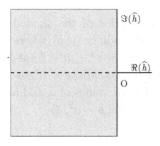

$\Im(\widehat{h})$

$\Re(\widehat{h})$

O

Fig. 6.7 The region of absolute stability for the trapezoidal rule (shaded) and the interval of absolute stability $(-\infty, 0)$ (Broken line). The axes have no scale because the region is infinite

Thus, if the trapezoidal rule were to be applied to the problems in Examples 6.1 or 6.8, we would have absolute stability regardless of the size of h; this means that h can be chosen on grounds of accuracy without regard to stability.

The following lemma is useful when the stability polynomial is quadratic and $\widehat{h}$ is real.

Lemma 6.10 (Jury Conditions)

The quadratic polynomial $q(r) = r^2 + ar + b$, where a and b are both real parameters, will satisfy the strict root condition of Definition 5.4 if, and only if,

$$\text{(i) } b < 1, \quad \text{(ii) } 1 + a + b > 0, \quad \text{and} \quad \text{(iii) } 1 - a + b > 0.$$

These are often called Jury conditions [40] and they define the triangular region shown in Figure 6.8.

Proof

Using the quadratic formula, the roots of $q(r) = r^2 + ar + b$ are given by

$$r_1, r_2 = \tfrac{1}{2}\left(-a \pm \sqrt{a^2 - 4b}\right).$$

When $a^2 < 4b$ the roots form a complex conjugate pair and, since b is equal to the product of the roots, we find that $b = r_1 r_2 = |r_1|^2 = |r_2|^2$. Hence, the strict root condition holds if, and only if, $b < 1$. The inequality $a^2 < 4b$ also implies that (ii) and (iii) are satisfied (see Exercise 6.4).

In the case of real roots ($a^2 \geq 4b$), the root of largest magnitude R is

$$R = \max\{|r_1|, |r_2|\} = \tfrac{1}{2}\left(|a| + \sqrt{a^2 - 4b}\right).$$

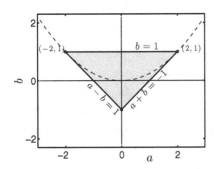

Fig. 6.8 The interior of the shaded triangle shows the points (a, b) where the polynomial $q(r) = r^2 + ar + b$ satisfies the strict root conditions. (Solutions are complex for values of a and b above the broken line: $b = a^2/4$)

This is an increasing function of $|a|$, and $R = 1$ when $|a| = 1 + b$. Hence, $0 \leq R < 1$ if, and only if, $0 \leq |a| < 1 + b$. Also, since the strict root condition implies that $|r_1 r_2| < 1$, it follows that $|b| < 1$ and condition (i) must hold.

Combining the results for real and complex roots, we find that the strict root condition is satisfied if, and only if, $|a| - 1 < b < 1$. $\square$

These conditions are equivalent to

$$q(0) < 1 \quad \text{and} \quad q(\pm 1) > 0,$$

which may be easier to remember. To apply this lemma to the stability polynomial (6.3) for a general two-step LMM it is first necessary to divide by the coefficient of r^2, so

$$a = \frac{\alpha_1 - \widehat{h}\beta_1}{1 - \widehat{h}\beta_2}, \qquad b = \frac{\alpha_0 - \widehat{h}\beta_0}{1 - \widehat{h}\beta_2}.$$

The denominators in these coefficients are necessarily positive for all $\widehat{h} \in \mathcal{R}_0$ (see Exercise 6.12), so the conditions $q(\pm 1) > 0$ can be replaced by $p(\pm 1) > 0$. For explicit LMMs, $\beta_2 = 0$ and $q(r)$ will coincide with the stability polynomial $p(r)$.

Example 6.11

Find the interval of absolute stability of the LMM:

$$x_{n+2} - x_{n+1} = h f_n.$$

The stability polynomial is quadratic in r:

$$p(r) = r^2 - r - \widehat{h}; \tag{6.4}$$

since $\widehat{h}$ is real, the coefficients of this polynomial are real, and we can use Lemma 6.10 to determine precisely when it satisfies the strict root condition. Moreover, $p(r) \equiv q(r)$, so the conditions for absolute stability are $p(\pm 1) > 0$ and $p(0) < 1$. We find

$$p(0) < 1 : -\widehat{h} < 1 \quad \Rightarrow \widehat{h} > -1,$$
$$p(1) > 0 : -\widehat{h} > 0 \quad \Rightarrow \widehat{h} < 0,$$
$$p(-1) > 0 : \ 2 - \widehat{h} > 0 \Rightarrow \widehat{h} < 2.$$

In order to satisfy all three inequalities, we must have $-1 < \widehat{h} < 0$. So the interval of absolute stability is $(-1, 0)$. $\square$

Example 6.12

Find the interval of absolute stability of the mid-point rule: $x_{n+2}-x_n = 2hf_{n+1}$.

In this case

$$p(r) = r^2 - 2\widehat{h}r - 1,$$

whose roots are

$$r_+ = \widehat{h} + \sqrt{1+\widehat{h}^2}, \qquad r_- = \widehat{h} - \sqrt{1+\widehat{h}^2},$$

and it should be immediately obvious that $|r_-| > 1$ when $\widehat{h} < 0$, so the method can never be absolutely stable. To see the consequences of this, the method is applied to the IVP $x'(t) = -8x(t)$ with $x(0) = 1$. The results are shown in Figure 6.9 with $h = 1/40$ ($\widehat{h} = 1/5$) and $h = 1/120$ ($\widehat{h} = 1/15$) and the additional starting value provided by the exact solution at $t = t_1$. The solutions of the mid-point rule are given by (the details are left to Exercise 6.13)

$$x_n = Ar_+^n + Br_-^n, \tag{6.5}$$

where the constants A and B are chosen to satisfy the starting conditions $x_0 = 1$, $x_1 = e^{\widehat{h}}$. It can be shown that

$$r_+ = e^{\widehat{h}} + \mathcal{O}(\widehat{h}^3) \quad \text{and} \quad r_- = -e^{-\widehat{h}} + \mathcal{O}(\widehat{h}^3),$$

so that $r_+^n = e^{\lambda t^*} + \mathcal{O}(h^2)$ and $r_-^n = (-1)^n e^{-\lambda t^*} + \mathcal{O}(h^2)$ at $t^* = nh$. The first of these approximates the corresponding term in the exact solution, $e^{\lambda t^*}$, while r_- has no such role—for this reason it is usually referred to as a *spurious root*: the ODE is of first order but the difference equation that approximates it is of second order. On solving for A and B and expanding the results in a Maclaurin series in $\widehat{h}$ we find

$$A = 1 + \mathcal{O}(h^2), \qquad B = -\tfrac{1}{12}h^3 + \mathcal{O}(h^4).$$

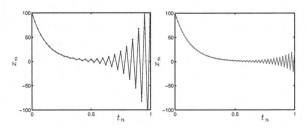

Fig. 6.9 Numerical solution by the mid-point rule for Example 6.12 with $h = 1/40$ (left) and $h = 1/120$ (right)

Hence, the first term Ar_+^n in the solution (6.5) approximates the exact solution to within $\mathcal{O}(h^2)$ while the second term satisfies

$$Br_-^n = -\tfrac{1}{12}h^3(-1)^n e^{-\lambda t^*} + \mathcal{O}(h^4).$$

It is this term that causes problems: it is exponentially growing when $\Re(\lambda) < 0$ (and the exact solution is exponentially decaying) and the factor $(-1)^n$ causes it to alternate in sign on consecutive steps (producing the oscillations evident in Figure 6.9). On a positive note, it has an amplitude proportional to h^3, so becomes negligible compared with the dominant $\mathcal{O}(h^2)$ term in the GE when h is sufficiently small.

Other Nyström and Milne–Simpson methods have similar properties, so cannot be recommended for solving problems with damping ($\Re(\lambda) < 0$). However, it is a different story if λ is purely imaginary (oscillatory problems)—see Section 7.3. □

6.3 The Boundary Locus Method

It is, in general, quite difficult to determine the region of absolute stability of an LMM since we have to decide, for each $\widehat{h} \in \mathbb{C}$, whether the roots of the stability polynomial satisfy the strict root condition ($|r| < 1$). It is more attractive to look for the boundary of the region, because at least one of the roots of $p(r)$ on the boundary has modulus $|r| = 1$. Thus, the boundary is a subset of the points $\widehat{h} \in \mathbb{C}$ for which $r = e^{is}$, where $s \in \mathbb{R}$. Substituting $r = e^{is}$ into the stability polynomial and solving for $\widehat{h}$ we obtain $\widehat{h} = \widehat{h}(s)$ and plotting the locus in the complex plane gives a curve, part of which will be the required boundary. We illustrate the process with an example that contains most of the important features.

Example 6.13

Use the boundary locus method to determine the boundary of the region of absolute stability of the LMM $x_{n+2} - x_{n+1} = hf_n$ (see Example 6.11).

With $r = e^{is}$ the stability polynomial (6.4) gives $\widehat{h} = r^2 - r$, and so

$$\widehat{h}(s) = e^{2is} - e^{is} = [\cos(2s) - \cos(s)] + i[\sin(2s) - \sin(s)].$$

Plotting the locus of the points $\widehat{x}(s) = \cos(2s) - \cos(s)$, $\widehat{y}(s) = \sin(2s) - \sin(s)$ for $0 \le s < 2\pi$ we obtain the curve shown in Figure 6.10, which divides the plane into three subregions—it remains to decide which subregion is the region

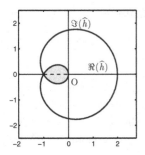

Fig. 6.10 Stability region $\mathcal{R}$ for Example 6.13 (shaded). The solid curve is the locus of points where the stability polynomial $p(r)$ has at least one root r with $|r| = 1$

of absolute stability. We need only test one point in each subregion: if the roots at that point satisfy the strict root condition then the point lies in the region of absolute stability, otherwise it does not lie in the region. This is done in the table below.

$\widehat{h}$	$p(r)$	Roots	$|r|$	Absolutely stable
$-\frac{1}{2}$	$r^2 - r + \frac{1}{2}$	$r = (1 \pm \mathrm{i})/2$	$|r| = 1/\sqrt{2} < 1$	yes
1	$r^2 - r - 1$	$r = (1 \pm \sqrt{5})/2$	$|r| > 1$	no
-2	$r^2 - r + 2$	$r = (1 \pm \mathrm{i}\sqrt{7})/2$	$|r| > 1$	no

We conclude that the region of absolute stability is the shaded region in Figure 6.10. The curve intersects itself when $\Im(\widehat{h}(s)) = 0$. This is easily shown to occur when $\cos s = \frac{1}{2}$, so $\sin s = \frac{1}{2}\sqrt{3}$ and $\widehat{h} = -1$. The interval of absolute stability is, therefore, $(-1, 0)$, in agreement with Example 6.11. $\qquad\square$

6.4 A-stability

Some LMMs (the trapezoidal rule is one example) applied to the model problem (6.1) have the satisfying property that $x_n \to 0$ as $n \to \infty$ whenever $x(t) \to 0$ as $t \to \infty$ regardless of the size of h. This is sufficiently important to give the set of all such methods a name:

Definition 6.14 (A-Stability)

A numerical method is said to be *A-stable* if its region of absolute stability $\mathcal{R}$ includes the entire left half plane ($\Re(\widehat{h}) < 0$).

This is a severe requirement, as evidenced by the following theorem.

Theorem 6.15 (Dahlquist's Second Barrier Theorem)

1. There is no A-stable explicit LMM.

2. An A-stable (implicit) LMM cannot have order $p > 2$.

3. The order-two *A-stable* LMM with scaled error constant $(C_{p+1}/\sigma(1))$ of smallest magnitude is the trapezoidal rule.

Proof

See Hairer and Wanner [29].   □

We can relax our requirements when λ is real.

Definition 6.16 (A_0-Stability)

A numerical method is said to be A_0-stable if its interval of absolute stability includes the entire left real axis $(\Re(\widehat{h}) < 0,\ \Im(\widehat{h}) = 0)$.

As a parting remark, we observe that A-stability has been defined in terms of the simple linear differential equation $x'(t) = \lambda x(t)$. What is remarkable (and not fully understood) is why methods which are A-stable generally outperform other methods on more general non-linear problems.

EXERCISES

6.1.** By writing $\widehat{h} = 2X + 2iY$ in Example 6.9 prove that

$$|r_1|^2 - 1 = \frac{4X}{(1 - X)^2 + Y^2}$$

and deduce that $|r_1| < 1$ for all $\Re(\widehat{h}) < 0$.

What can be concluded about the interval of absolute stability of the trapezoidal rule?

6.2.** Prove that the region of absolute stability of the backward Euler method $x_{n+1} = x_n + hf_{n+1}$ is given by $|1 - \widehat{h}| > 1$. By writing $\widehat{h} = \widehat{x} + i\widehat{y}$ show that this corresponds to the exterior of the circle $(\widehat{x} - 1)^2 + \widehat{y}^2 = 1$. Sketch a diagram analogous to Figure 6.6.

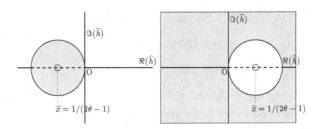

Fig. 6.11 The region of absolute stability for the θ-method (shaded) and the interval of absolute stability (broken line). Left: $0 \leq \theta < \frac{1}{2}$; right: $\frac{1}{2} < \theta \leq 1$. See Exercise 6.3.

6.3.** Determine the region of absolute stability of the θ-method

$$x_{n+1} - x_n = h(\theta f_{n+1} + (1-\theta)f_n).$$

By writing $\widehat{h} = \widehat{x} + i\widehat{y}$ show that this corresponds to the exterior of the circle

$$\widehat{x}^2 + \frac{2}{2\theta - 1}\widehat{x} + \widehat{y}^2 = 0$$

if $\frac{1}{2} < \theta \leq 1$ and to the interior of the circle if $0 \leq \theta < \frac{1}{2}$. See Figure 6.11. What happens at $\theta = \frac{1}{2}$?

6.4.* Verify the identity

$$a^2 - 4b = (|a| - 2)^2 - 4(1 + b - |a|)$$

for any real numbers a, b and deduce that $b > |a| - 1$ whenever $a^2 < 4b$, as required in the proof of Lemma 6.10.

6.5.* Find the interval of absolute stability of the LMM

$$x_{n+2} - x_n = \frac{1}{2}h(f_{n+1} + 3f_n).$$

6.6.* Find the interval of absolute stability of the two-step Adams–Bashforth method (AB(2))

$$x_{n+2} - x_{n+1} = \frac{1}{2}h(3f_{n+1} - f_n).$$

6.7.** Consider the one-parameter family of LMMs

$$x_{n+2} - 4\theta x_{n+1} - (1 - 4\theta)x_n = h\big((1-\theta)f_{n+2} + (1 - 3\theta)f_n\big)$$

for solving the ODE $x'(t) = f(t, x)$, where the notation is standard.

(a) Determine the error constant for this family of methods and identify the method of highest order. What is its error constant?

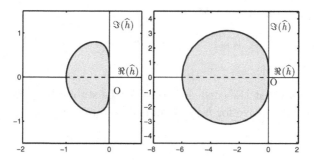

Fig. 6.12 Stability regions $\mathcal{R}$ for Exercise 6.9. One belongs to the Adams–Bashforth method AB(2) and the other to the Adams–Moulton method AM(2)

(b) For what values of θ are members of this family of methods convergent?

Is the method of highest order convergent? Explain your answer carefully.

(c) Prove that the method is A_0-stable when $\theta = 1/4$.

6.8.*** What is the stability polynomial of the LMM

$$x_{n+2} - (1+a)x_{n+1} + ax_n =$$
$$\tfrac{1}{12}h[(5+a)f_{n+2} + 8(1-a)f_{n+1} - (1+5a)f_n].$$

– Why must the condition $-1 \le a < 1$ be stipulated?

– Assuming that $\widehat{h} \in \mathbb{R}$, show that the discriminant of the (quadratic) stability polynomial is strictly positive for all values of $\widehat{h}$ and a. What information does this give regarding its roots?

– Deduce that the interval of absolute stability is $(-6\frac{1+a}{1-a}, 0)$.

6.9.** Shown in Figure 6.12 are the stability regions of the AB(2) (see Exercise 6.6) and the Adams–Moulton method (AM(2))

$$x_{n+2} - x_{n+1} = \tfrac{1}{12}h(5f_{n+2} + 8f_{n+1} - f_n). \tag{6.6}$$

Use the boundary locus method to ascertain which region belongs to which method.

6.10.* Apply the boundary locus method to the mid-point rule (see Example 6.12) to show that
$$\widehat{h}(s) = \mathrm{i}\sin s$$

so that the region of absolute stability consists of that part of the imaginary axis between $-\mathrm{i}$ and i.

6.11.** Show that the LMM

$$x_{n+2} - x_{n+1} = \tfrac{1}{4}h(f_{n+2} + 2f_{n+1} + f_n)$$

is A_0-stable.

6.12.** Prove that the coefficient of r^2 in the stability polynomial (6.3) is always positive for $\widehat{h}$ in the interval of absolute stability; i.e.

$$1 - \widehat{h}\beta_2 > 0 \text{ for all } \widehat{h} \in \mathcal{R}_0.$$

6.13.*** Complete the details in Example 6.12. Show, in particular, that $A = \tfrac{1}{2}(1 + a)$ and $B = \tfrac{1}{2}(1 - a)$, where

$$a = \frac{e^{\widehat{h}} - \widehat{h}}{\sqrt{1 + \widehat{h}^2}} = 1 + \tfrac{1}{6}\widehat{h}^3 + \mathcal{O}(\widehat{h}^4).$$

Deduce that $A = 1 + \mathcal{O}(\widehat{h}^3)$ and $B = \mathcal{O}(\widehat{h}^3)$.

6.14.** Find the interval of absolute stability of each member of the convergent family of LMMs of the form (see Exercise 5.11)

$$x_{n+2} - x_n = h\left(\beta_1 f_{n+1} + \beta_0 f_n\right).$$

Why can no member of the family be identified as having the largest such interval?

6.15.*** Show that all convergent members of the family of methods

$$x_{n+2} + (\theta - 2)x_{n+1} + (1 - \theta)x_n = \tfrac{1}{4}h\left[(6 + \theta)f_{n+2} + 3(\theta - 2)f_n\right],$$

parameterized by θ, are also A_0-stable.

6.16.** Show that the method

$$2x_{n+2} - 3x_{n+1} + x_n = \tfrac{1}{2}h\left(4f_{n+2} - 3f_{n+1} + f_n\right)$$

is A_0-stable.

6.17.*** Consider the family of LMMs

$$x_{n+2} - 2ax_{n+1} + (2a - 1)x_n = h\left[af_{n+2} + (2 - 3a)f_{n+1}\right],$$

where a is a parameter.

(a) What are its first and second characteristic polynomials?

(b) When is the method consistent?

(c) Under what conditions is it zero-stable?

(d) When is the method convergent?

(e) What is its order? What is the error constant?

(f) Are there any members of the family that are A_0-stable?

(g) What conclusions can you draw concerning the backward differentiation formula (BDF(2))

$$3x_{n+2} - 4x_{n+1} + x_n = 2hf_{n+2}?$$

(h) Verify that all three statements of Dahlquist's second barrier theorem hold for this family of methods.

6.18.** Show that the method $2x_{n+2} - x_{n+1} - x_n = 3hf_{n+2}$ is A_0-stable.

Use the boundary locus method to show that, on the boundary of the region of stability, $\widehat{h}(s) = (2 - e^{-is} - e^{-2is})/3$. Deduce that $\Re(\widehat{h}(s)) \geq 0$ for all s and conclude that the method is A-stable.

6.19.* Use Lemma 6.10 to derive necessary and sufficient conditions on the (real) coefficients (a, b) for the roots of the polynomial $q(r) = r^2 + ar + b$ to satisfy the root condition. Pay particular attention to double roots having modulus equal to one.

6.20.*** Suppose that a convergent LMM has a cubic characteristic polynomial

$$\rho(r) = r^3 + ar^2 + br + c.$$

Prove that

(a) it can be factorized as $\rho(r) = (r - 1)(r^2 + (1 + a)r - c)$;

(b) the coefficients a and c must satisfy

$$2 + a - c > 0, \quad a + c \leq 0, \quad 1 + c \geq 0$$

while excluding the point $(a, c) = (1, -1)$.

[Hint: use Lemma 6.10, paying particular attention to roots of $\rho(r)$ with $|r| = 1$.]

6.21.*** The *composite Euler method* uses Euler's method with a step size h_0 on even-numbered steps and a step size h_1 on odd-numbered steps, so

$$x_{2m+1} = x_{2m} + h_0 f_{2m} \quad \text{and} \quad x_{2m+2} = x_{2m+1} + h_1 f_{2m+1},$$

where $h_0 = (1 - \gamma)h$, $h_1 = (1 + \gamma)h$ $(0 \leq \gamma < 1)$ so a time interval of $2h$ is covered in two consecutive steps. The interval of absolute stability of this composite method can be deduced by applying it to the ODE $x'(t) = \lambda x(t)$ and examining the ratio $R(\widehat{h}) = x_{2m+2}/x_{2m}$. Show that $R(\widehat{h}) = (1 + \widehat{h})^2 - \gamma^2 \widehat{h}^2$.

By determining the global minimum of the function $R(\widehat{h})$, prove that $-1 < R(\widehat{h}) < 1$ if, and only if, $\widehat{h}$ lies in the interval of absolute stability given by

$$-\frac{2}{1 - \gamma^2} < \widehat{h} < 0.$$

Deduce that the largest interval of absolute stability is $\widehat{h} \in (-4, 0)$ and occurs when $\gamma = 1/\sqrt{2}$. Hence, the composite method may be used in a stable manner with a value of h that is twice as large as the standard Euler method. The region of absolute stability with $\gamma = 0$ is the same as for Euler's method (Figure 6.8); when $\gamma = 3/5$ and $1/\sqrt{2}$ it is as shown in Figure 6.13.

6.22.** Suppose that $\gamma = 1/\sqrt{2}$ and $\lambda = -8 + i$ in the previous exercise. Show that the method will be absolutely stable when $h = 1/8$ and $3/8$ but not when $h = 1/4$ and $1/2$. Relate these results to the region of absolute stability shown in Figure 6.13. See also Exercise 10.14.

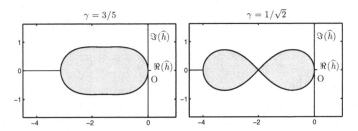

Fig. 6.13 The region of absolute stability for the composite Euler method for Exercise 6.21 when $\gamma = 3/5$ and $1/\sqrt{2}$

7

Linear Multistep Methods—IV: Systems of ODEs

In this chapter we describe the use of LMMs to solve systems of ODEs and show how the notion of absolute stability can be generalized to such problems. We begin with an example.

Example 7.1

Use the LMM (AB(2))

$$x_{n+2} - x_{n+1} = \tfrac{1}{2}h(3f_{n+1} - f_n)$$

to compute the solution at $t = 0.2$ of the IVP

$$u'(t) = -tu(t)v(t),$$
$$v'(t) = -u^2(t),$$

with $u = 1$, $v = 2$ at $t = 0$. Use $h = 0.1$ and Euler's method to calculate u_1 and v_1.

We begin by using the differential equations to calculate $u'(0) = 0$ and $v'(1) = -1$ so we have the starting values

$$t_0 = 0, \quad u_0 = 1, \quad v_0 = 2, \quad u'_0 = 0, \quad v'_0 = -1.$$

D.F. Griffiths, D.J. Higham, *Numerical Methods for Ordinary Differential Equations*,
Springer Undergraduate Mathematics Series, DOI 10.1007/978-0-85729-148-6_7,
© Springer-Verlag London Limited 2010

Then, applying Euler's method (see Section 2.5) to each of the individual ODEs:

$$n = 0 : \quad t_1 = 0.1,$$
$$u_1 = u_0 + hu_0' = 1,$$
$$v_1 = v_0 + hv_0' = 1.9,$$

with which we can compute $u_1' = -t_1 u_1 v_1 = -0.19$ and $v_1' = -u_1^2 = -1$.

When the given LMM is applied to both the u and v differential equations it is seen that we have to calculate, for each $n = 0, 1, 2, \ldots$,

$$t_{n+2} = t_{n+1} + h,$$
$$u_{n+2} = u_{n+1} + \tfrac{1}{2}h(3u_{n+1}' - u_n'),$$
$$v_{n+2} = v_{n+1} + \tfrac{1}{2}h(3v_{n+1}' - v_n'),$$
$$u_{n+2}' = -t_{n+2}u_{n+2}v_{n+2},$$
$$v_{n+2}' = -u_{n+2}^2.$$

So,

$$n = 0 : \quad t_2 = t_1 + h = 0.2,$$
$$u_2 = u_1 + \tfrac{1}{2}h(3u_1' - u_0') = 0.9715,$$
$$v_2 = v_1 + \tfrac{1}{2}h(3v_1' - v_0') = 1.8,$$
$$u_2' = -t_2 u_2 v_2 = -0.19,$$
$$v_2' = -u_2^2 = -1.0.$$

The computations of the first few steps are summarized in Table 7.1.

More generally, for the system of ODEs written in vector form,

$$\boldsymbol{u}'(t) = \boldsymbol{f}(t, \boldsymbol{u}(t)), \qquad t > t_0,$$

with $\boldsymbol{u}(t_0) = \boldsymbol{u}_0$, the first step is computed by Euler's method , after which

$$t_{n+2} = t_{n+1} + h,$$
$$\boldsymbol{u}_{n+2} = \boldsymbol{u}_{n+1} + \tfrac{1}{2}h(3\boldsymbol{u}_{n+1}' - \boldsymbol{u}_n'),$$
$$\boldsymbol{u}_{n+2}' = \boldsymbol{f}(t_{n+2}, \boldsymbol{u}_{n+2}),$$

for $n = 0, 1, 2, \ldots$.   □

n	t_n	u_n	v_n	u_n'	v_n'	
0	0	1.0000	2.0000	0	−1.0000	Initial data
1	0.1000	1.0000	1.9000	−0.1900	−1.0000	Euler's Method
2	0.2000	0.9715	1.8000	−0.1900	−1.0000	AB(2)
3	0.3000	0.9525	1.7000	−0.3497	−0.9438	...

Table 7.1 Numerical solutions for Example 7.1

7.1 Absolute Stability for Systems

The notion of absolute stability for systems of ODEs requires us to apply our LMM to the *model* problem involving a system of m first-order linear ODEs with constant coefficients:

$$u'(t) = Au(t), \tag{7.1}$$

where $u(t)$ is an m-dimensional vector $(u(t) \in \mathbb{R}^m)$ and A is a constant $m \times m$ matrix $(A \in \mathbb{R}^{m \times m})$. We first recall some aspects of ODE theory.

Diagonalization of ODEs. In order to understand how solutions of the ODE system behave we carry out a "diagonalization" process. We assume that A is a diagonalizable matrix, i.e. it has m linearly independent eigenvectors $v_1, \ldots, v_m$ with corresponding eigenvalues $\lambda_1, \ldots, \lambda_m$:

$$Av_j = \lambda_j v_j.$$

Under these circumstances, there exists a (possibly complex) nonsingular matrix V whose columns are the vectors $v_1, \ldots, v_m$, such that

$$V^{-1}AV = \Lambda,$$

where Λ is the $m \times m$ diagonal matrix with entries $\lambda_1, \ldots, \lambda_m$ on the diagonal. Defining $u(t) = Vx(t)$, then $x(t)$ satisfies the differential equation

$$x'(t) = \Lambda x(t), \tag{7.2}$$

a typical component of which is

$$x'(t) = \lambda x(t), \tag{7.3}$$

where λ is an eigenvalue of A. The solution of the linear system (7.1) has been reduced to solving a collection of scalar problems of the type studied in Section 6.2, one for each eigenvalue of A. Since $x(t)$ and $u(t)$ are connected through a fixed linear transformation, they have the same long-term behaviour. Hence, the scalar problems (7.3) tell us everything we need to know. This is illustrated in the next example and formalized in the theorem that follows.

Example 7.2

Determine the general solution of the system (7.1) when

$$A = \begin{bmatrix} 1 & 3 \\ -2 & -4 \end{bmatrix}$$

and examine the behaviour of solutions as $t \to \infty$.

The matrix A may be diagonalized using

$$V = \begin{bmatrix} 3 & -1 \\ -2 & 1 \end{bmatrix}$$

into the form

$$V^{-1}AV = \begin{bmatrix} -1 & 0 \\ 0 & -2 \end{bmatrix}.$$

This shows that A has eigenvalues $\lambda_1 = -1$ and $\lambda_2 = -2$, with corresponding eigenvectors

$$v_1 = \begin{bmatrix} 3 \\ -2 \end{bmatrix}, \qquad v_2 = \begin{bmatrix} -1 \\ 1 \end{bmatrix}.$$

In some circumstances it would be convenient to work with the normalized eigenvectors $\frac{1}{\sqrt{13}}v_1$ and $\frac{1}{\sqrt{2}}v_2$, but in our context normalizing is not helpful. Using $u'(t) = Au(t)$, the new variables $x(t) = V^{-1}u(t)$ satisfy

$$x'(t) = V^{-1}u'(t) = V^{-1}Au(t) = V^{-1}AVx(t) = \begin{bmatrix} -1 & 0 \\ 0 & -2 \end{bmatrix} x(t).$$

We have now uncoupled the system into the two scalar problems $x'(t) = -x(t)$ and $y'(t) = -2y(t)$, where $x(t) = [x(t), y(t)]^{\mathrm{T}}$. These have general solutions $x(t) = A\,\mathrm{e}^{-t}$ and $y(t) = B\,\mathrm{e}^{-2t}$, where the constants A and B depend on the initial data. It follows that $u(t)$ has the general form

$$u(t) = Vx(t) = \begin{bmatrix} 3 & -1 \\ -2 & 1 \end{bmatrix} \begin{bmatrix} A\,\mathrm{e}^{-t} \\ B\,\mathrm{e}^{-2t} \end{bmatrix} = A\,\mathrm{e}^{-t} \begin{bmatrix} 3 \\ -2 \end{bmatrix} + B\,\mathrm{e}^{-2t} \begin{bmatrix} -1 \\ 1 \end{bmatrix}.$$

It is now obvious that $u(t) \to 0$ as $t \to \infty$, and it is clear from the derivation that this property follows directly from the nature of the two eigenvalues. □

Absolute stability is concerned with solutions of unforced ODEs that tend to zero as $t \to \infty$ and these are characterized in the following theorem.

Theorem 7.3

If A is a diagonalizable matrix having eigenvalues $\lambda_1, \ldots, \lambda_m$, then the solutions of $u'(t) = Au(t)$ tend to zero as $t \to \infty$ for all choices of initial conditions if, and only if, $\Re(\lambda_j) < 0$ for each $j = 1, 2, \ldots, m$. ($\Re(\lambda)$ denotes the real part of λ.)

For a proof see Braun [5, Section 4.2], Nagle et al. [57, Section 12.7] or O'Malley [59, Section 5.3].

Diagonalization of LMMs. Applying the general two-step LMM

$$x_{n+2} + \alpha_1 x_{n+1} + \alpha_0 x_n = h(\beta_2 f_{n+2} + \beta_1 f_{n+1} + \beta_0 f_n) \qquad (7.4)$$

to the system $u'(t) = Au(t)$ we obtain (since $f_n = Au_n$)

$$u_{n+2} + \alpha_1 u_{n+1} + \alpha_0 u_n = hA(\beta_2 u_{n+2} + \beta_1 u_{n+1} + \beta_0 u_n). \qquad (7.5)$$

Following the treatment of scalar ODEs, we define $u_{n+j} = V x_{n+j}$ for each n and each j. Then, (7.5) becomes, on multiplying by V^{-1},

$$V^{-1} u_{n+2} + \alpha_1 V^{-1} u_{n+1} + \alpha_0 V^{-1} u_n$$
$$= hV^{-1}A(\beta_2 u_{n+2} + \beta_1 u_{n+1} + \beta_0 u_n)$$
$$= hV^{-1}AV(\beta_2 V^{-1} u_{n+2} + \beta_1 V^{-1} u_{n+1} + \beta_0 V^{-1} u_n),$$

which simplifies to

$$x_{n+2} + \alpha_1 x_{n+1} + \alpha_0 x_n = h\Lambda(\beta_2 x_{n+2} + \beta_1 x_{n+1} + \beta_0 x_n).$$

This is precisely the recurrence that arises when we apply the LMM directly to the diagonalized system of ODEs (7.2). The components are now uncoupled and, if we write x_n to denote a typical component of x_n, we find

$$x_{n+2} + \alpha_1 x_{n+1} + \alpha_0 x_n = h\lambda(\beta_2 x_{n+2} + \beta_1 x_{n+1} + \beta_0 x_n), \qquad (7.6)$$

in which λ is a typical eigenvalue of A. This is the same equation (6.2) that was obtained when the general two-step LMM (7.4) was applied to the scalar ODE (7.3).

We have shown that the two processes

1. apply the LMM

2. diagonalise A

commute—the same result is obtained regardless of the order in which the operations are carried out. This can be illustrated by the "commutative diagram"

$$
\begin{array}{ccc}
u' = Au & \xrightarrow{\text{Apply LMM}} & (7.5) \\
\downarrow{\text{Diagonalize}} & & \downarrow{\text{Diagonalize}} \\
x' = \Lambda x & \xrightarrow{\text{Apply LMM}} & (7.6)
\end{array}
$$

in which the route taken from top left to bottom right is immaterial.[1]

We now return to the question of absolute stability—the following definition is simply a rephrasing of Definition 6.3 to accommodate systems of ODEs and k-step methods.

---

[1] In practice, though, the constants hidden inside the arrows can be important. See Trefethen and coworkers [35, 67] for details of the fascinating topic of *pseudoeigenvalues*.

Definition 7.4 (Absolute Stability)

Suppose that all solutions of $u'(t) = Au(t)$ tend to zero as $t \to \infty$ for all choices of initial condition u_0. A k-step numerical method with a given stepsize h applied to such a system is said to be *absolutely stable* if all its solutions tend to zero as $n \to \infty$ for all choices of starting values $u_0, u_1, \ldots, u_{k-1}$.

Using the diagonalization trick that produced (7.2) and (7.3), we can appeal to the scalar analysis from Chapter 6. We know that $u_n \to 0$ if, and only if, $x_n \to 0$, and hence we simply require that all solutions to (7.6) tend to zero for every eigenvalue λ of A. Thus

An LMM is absolutely stable for the diagonalizable system $u'(t) = Au(t)$ if $\lambda h \in \mathcal{R}$ (the region of absolute stability) for every eigenvalue λ of A.

In other words, to analyse the behaviour of any LMM applied to (7.1), we need only consider its application to the scalar problem (7.3) and our results can be transferred to systems of ODEs simply by interpreting λ as any eigenvalue of A.

Example 7.5

What is the largest step size h allowed by absolute stability when the system

$$u'(t) = -11u(t) + 100v(t), \qquad v'(t) = u(t) - 11v(t) \tag{7.7}$$

is solved using Euler's method?

Euler's method is, in this case,

$$u_{n+1} = u_n + h\big(-11u_n + 100v_n\big),$$
$$v_{n+1} = v_n + h\big(u_n - 11v_n\big).$$

The system of ODEs may be written in matrix-vector form $u'(t) = Au(t)$ with coefficient matrix

$$A = \begin{bmatrix} -11 & 100 \\ 1 & -11 \end{bmatrix},$$

which has eigenvalues -1 and -21. Since these are real we use the interval of absolute stability of Euler's method (see Example 6.7), which requires $h\lambda \in (-2, 0)$. Therefore, h must satisfy

$$-2 < -h < 0 \qquad \text{and} \qquad -2 < -21h < 0,$$

i.e. $0 < h < \frac{2}{21} \approx 0.0952$.

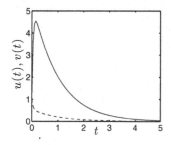

Fig. 7.1 The components $u(t)$ (solid curve) and $v(t)$ (dashed curve) of the solution of the IVP of Example 7.5

The exact solution of the ODEs with initial values $u(0) = v(0) = 1$ is shown in Figure 7.1. There is a "rapid initial transient" as one component grows to its maximum at a time $t = 0.14$, after which it decays more gradually to zero. The fast transient is governed by the eigenvalue $\lambda = -21$ and the slow decay by $\lambda = -1$, the exact solution being (see Exercise 7.1)

$$\begin{bmatrix} u \\ v \end{bmatrix} = \frac{11}{20} \begin{bmatrix} 10 \\ 1 \end{bmatrix} e^{-t} - \frac{9}{20} \begin{bmatrix} 10 \\ -1 \end{bmatrix} e^{-21t}. \tag{7.8}$$

Results for Euler's method with various step sizes are shown in Figure 7.2.

$h = 0.0962$, which is 1% over the stability limit $h = 2/21$ (Figure 7.2 left). Although the amplitude of the solution grows only slowly, the rapid oscillations are a tell-tale sign of instability.

$h = 0.0905$, which is 5% under the limit (Figure 7.2 middle). Although the solution tends to zero (because we are within the limit of absolute stability) there continue to be strong oscillations and the solution does not begin to be accurate until t nears the end of the interval of integration.

$h = 0.0476$, which is 50% of the limit (Figure 7.2 right). We now have a smooth

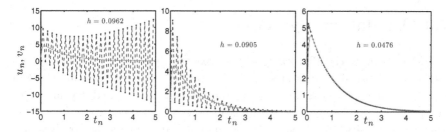

Fig. 7.2 The $u(t)$ component in Example 7.5 (solid line) and the corresponding component of the numerical solution u_n by Euler's method (dashed line)

solution ($u(t)$ is the solid line and $v(t)$ the dashed line).[2] The numerical solution is indistinguishable from the exact solution (thin solid line) for $t > 0.2$.

7.2 Stiff Systems

Examples 6.1 and 7.5 show that in certain types of system we have to use a small step size in order to produce an absolutely stable solution when, on grounds of accuracy, we would have expected to have been able to use a much larger value of h. Linear problems of this kind are characterized by matrices whose eigenvalues have negative real parts (so that $\boldsymbol{u}(t) \to 0$ as $t \to \infty$), but some have small absolute values while others are very large. That is, the ratio

$$\frac{\max_j -\Re(\lambda_j)}{\min_j -\Re(\lambda_j)}$$

may be extremely large (a ratio of 10^6 is not uncommon). We say that such problems are *stiff*. If we apply a method whose region of absolute stability is bounded, we find that the step length is restricted by the most negative eigenvalue, while the long-term solution $\boldsymbol{u}(t)$ will be dominated by the least negative eigenvalues. Therefore, very many time steps must be taken to compute the solution over a moderately long time. A by-product is that the numerical solution may be much more accurate than is needed.

The use of an A-stable method in such circumstances allows h to be chosen simply on grounds of accuracy, with no regard for stability. Thus, A-stable methods are particularly important for stiff systems.

7.3 Oscillatory Systems

Although all unforced physical processes exhibit some level of damping (and their mathematical models should, therefore, be solved by absolutely stable methods), some processes are best regarded as being undamped. Examples are (a) the motion of the planets, (b) inviscid flow of high-speed gases (where the effects of viscosity of a fluid may be ignored), and (c) molecular dynamics (which simulates the interactions of atoms and molecules). The Lotka–Volterra

---

[2]To avoid oscillations, the roots of the stability polynomial should satisfy $0 < r < 1$ (rather than $-1 < r < 1$) and, for Euler's method, this leads to $\lambda h \in (-1, 0)$.

equations (Example 1.3) modelling a naive predator-prey situation also have no damping.

Allowing complex-valued functions, the simplest oscillator is described by

$$x'(t) = \mathrm{i}\omega x(t),$$

which is a special case of the equation $x'(t) = \lambda x(t)$ we have met several times (see (6.1) and (7.3), for instance) with $\lambda = \mathrm{i}\omega$, an imaginary number. The general solution of this equation is $x(t) = c\,\mathrm{e}^{\mathrm{i}\omega t}$, where c is an arbitrary constant. Taking the absolute value, we have

$$|x(t)| = |c|,$$

and the motion in the complex plane is circular with a radius dictated by the starting condition. Numerical methods can be applied to complex problems, but we prefer to use the equivalent real system (see Exercise 1.6). The next example illustrates the unsuitability of both forward and backward Euler methods for solving oscillatory problems. A particularly large step size is chosen to exaggerate the effects so that they are more easily visualized.

Example 7.6

Use Euler's method, the trapezoidal rule and the backward Euler method with $h = 0.5$ to solve the IVP

$$u'(t) = -v(t), \qquad v'(t) = u(t),$$
$$u(0) = 1, \qquad v(0) = 0.$$

We calculate

$$\frac{\mathrm{d}}{\mathrm{d}t}\left(u^2(t) + v^2(t)\right) = 2u(t)u'(t) + 2v(t)v'(t) = 0 \qquad (7.9)$$

so that the function $u^2(t) + v^2(t)$ remains constant in time: the motion in the u-v phase plane is circular. This example examines which of the one-step methods can reproduce this type of motion.

Euler's method: $u_{n+1} = u_n - hv_n$ and $v_{n+1} = v_n + hu_n$ for $n = 0, 1, \ldots$ with $u_0 = 1$ and $v_0 = 0$.

The first three steps are shown on the left of Figure 7.3 ($\circ$ symbols) and the numerical solution clearly spirals outwards. At the nth step the motion generated by Euler's method is tangential to the circle with radius $\sqrt{u_n^2 + v_n^2}$ (these circles are shown as dotted curves). An easy calculation shows that

$$u_{n+1}^2 + v_{n+1}^2 = (1 + h^2)(u_n^2 + v_n^2),$$

so that the distance to the origin increases by a factor $\sqrt{1 + h^2}$ at each step. Thus, Euler's method displays a weak form of instability—not sufficient to prevent convergence (were we to take the limit $h \to 0$) but strong enough to make it unsuitable for simulating, for instance, the motion of the planets.

Backward Euler method: $u_{n+1} = u_n - hv_{n+1}$ and $v_{n+1} = v_n + hu_{n+1}$ for $n = 0, 1, \ldots$ with $u_0 = 1$ and $v_0 = 0$. Thus,

$$\begin{bmatrix} 1 & h \\ -h & 1 \end{bmatrix} \begin{bmatrix} u_{n+1} \\ v_{n+1} \end{bmatrix} = \begin{bmatrix} u_n \\ v_n \end{bmatrix}$$

leading to

$$\begin{bmatrix} u_{n+1} \\ v_{n+1} \end{bmatrix} = \frac{1}{1 + h^2} \begin{bmatrix} 1 & h \\ -h & 1 \end{bmatrix} \begin{bmatrix} u_n \\ v_n \end{bmatrix}.$$

The first three steps are shown on the right of Figure 7.3 (∘ symbols) and the numerical solution clearly spirals inwards. At the nth step the motion is tangential to the circle with radius $\sqrt{u_{n+1}^2 + v_{n+1}^2}$. It can be shown that

$$u_{n+1}^2 + v_{n+1}^2 = \frac{1}{1 + h^2}(u_n^2 + v_n^2),$$

so that the distance to the origin decreases by a factor $1/\sqrt{1 + h^2}$ at each step. Thus, the backward Euler method applies too much damping at each step.

Trapezoidal rule: $u_{n+1} = u_n - \frac{1}{2}h(v_{n+1} + v_n)$ and $v_{n+1} = v_n + \frac{1}{2}h(u_{n+1} + u_n)$ for $n = 0, 1, \ldots$ with $u_0 = 1$ and $v_0 = 0$. Thus,

$$\begin{bmatrix} 1 & h/2 \\ -h/2 & 1 \end{bmatrix} \begin{bmatrix} u_{n+1} \\ v_{n+1} \end{bmatrix} = \begin{bmatrix} 1 & -h/2 \\ h/2 & 1 \end{bmatrix} \begin{bmatrix} u_n \\ v_n \end{bmatrix}$$

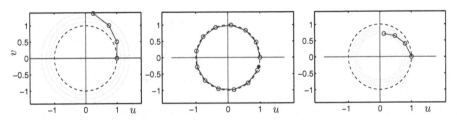

Fig. 7.3 Numerical solution of the system in Example 7.6 by Euler's method (left), trapezoidal rule (middle), and backward Euler (right), all using $h = 1/2$ and drawn in the u-v phase plane. The exact solution is shown by the dashed circle

leading to

$$\begin{bmatrix} u_{n+1} \\ v_{n+1} \end{bmatrix} = \frac{1}{4+h^2} \begin{bmatrix} 4-h^2 & -4h \\ 4h & 4-h^2 \end{bmatrix} \begin{bmatrix} u_n \\ v_n \end{bmatrix}. \tag{7.10}$$

The first 12 steps are shown in Figure 7.3 (centre: ∘ symbols) and the numerical solution appears to follow a circular motion: it can be confirmed algebraically that (7.10) implies that (see Exercise 7.7)

$$u_{n+1}^2 + v_{n+1}^2 = u_n^2 + v_n^2. \tag{7.11}$$

The trapezoidal rule faithfully computes the amplitude of the solution; it has, however, a second-order phase error—see Exercise 7.7. □

More generally, linear systems of higher dimension have the familiar structure

$$\boldsymbol{u}'(t) = A\boldsymbol{u}(t),$$

but, for oscillatory problems, the eigenvalues of A are imaginary numbers and so A is typically a skew-symmetric matrix. In Chapter 14 we consider general quadratic invariants, of which (7.9) is a special case.

7.4 Postscript

It would be wrong to give the impression that LMMs whose stability polynomials have all their roots strictly on the unit circle are the only, or even preferred, candidates for solving undamped problems.

When an absolutely stable numerical method is used to solve a damped problem (characterized by having eigenvalues with negative real parts) the LTE committed at one step is itself damped in subsequent steps. This can be seen most clearly in the expression (2.16) for the GE for Euler's method applied to $x'(t) = \lambda x(t)$—the effect at $t = t_n$ of the LTE, T_j, committed at the jth step is

$$(1+h\lambda)^{n-j}T_j,$$

with $|1+h\lambda| < 1$. A similar argument applies to the rounding error committed at the jth step.

In contrast, local errors committed in undamped LMMs can persist for all time (in periodic problems the LTE will also be periodic, so there will usually be some measure of cancellation of local errors over a period). There may, however, be some advantage in these situation in using LMMs that include a small amount of damping.

EXERCISES

7.1.* Confirm by direct differentiation that (7.8) solves the system of ODEs (7.7).

7.2.* Following the ideas in Example 7.2, investigate the $t \to \infty$ behaviour of solutions to (7.1) in the case where

$$A = \begin{bmatrix} 27 & -15 \\ 50 & -28 \end{bmatrix}.$$

We will start you off with the observation that

$$\begin{bmatrix} 27 & -15 \\ 50 & -28 \end{bmatrix} \begin{bmatrix} 3 \\ 5 \end{bmatrix} = \begin{bmatrix} 6 \\ 10 \end{bmatrix} \quad \text{and} \quad \begin{bmatrix} 27 & -15 \\ 50 & -28 \end{bmatrix} \begin{bmatrix} 1 \\ 2 \end{bmatrix} = \begin{bmatrix} -3 \\ -6 \end{bmatrix}.$$

7.3.* What condition must the step length h satisfy in order to achieve absolute stability when Euler's method is applied to the system $u'(t) = v(t)$, $v'(t) = -200u(t) - 20v(t)$?

7.4.** What is the largest value of h for which Euler's method is absolutely stable when applied to the system $\boldsymbol{u}'(t) = A\boldsymbol{u}(t)$ when

$$A = \begin{bmatrix} -1 & 1 \\ -1 & -1 \end{bmatrix}?$$

7.5.* Show that $0 < h < \frac{1}{4}$ is required for absolute stability when the IVP (see Equation (1.14))

$$u'(t) = -8(u(t) - v(t)), \qquad u(0) = 100,$$
$$v'(t) = -(v(t) - 5)/8, \qquad v(0) = 20$$

is solved using Euler's method.

The exact solution of the IVP is shown in Figure 1.9 (right) and the numerical solutions with $h = 1/3$, $1/5$, and $1/9$ in Figure 7.4. The behaviour of the u-component of the solution ($\circ$ and dashed lines) is almost identical to that for the equivalent scalar problem discussed in Example 6.1 (see Figure 6.2).

7.6.* Show that $u(t)$ in Example 7.6 satisfies the *second-order* ODE $u''(t) + u(t) = 0$ (known as the simple harmonic equation) with initial conditions $u(0) = 1$, $u'(0) = 0$.

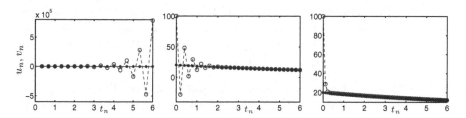

Fig. 7.4 Solution of the system in Exercise 7.5 by Euler's method with $h = 1/3$ (left), $h = 1/5$ (middle) and $h = 1/9$ (right). The u-component is depicted by ○/dashed lines and the v-component by ●/solid lines

7.7.*** Prove that (7.11) follows from (7.10).

In view of this relationship one may write $u_n = R\cos(\theta_n)$ and $v_n = R\sin(\theta_n)$, where $R^2 = u_0^2 + v_0^2$. Prove that

$$\tan(\theta_{n+1} - \theta_n) = \frac{h}{1 - h^2/4}$$

and, consequently, $\tan \frac{1}{2}(\theta_{n+1} - \theta_n) = \frac{1}{2}h$.

Use the Maclaurin expansion $\tan^{-1} z = z - \frac{1}{3}z^3 + \mathcal{O}(z^5)$ to show that the numerical solution rotates through an angle

$$\theta_{n+1} - \theta_n = h - \frac{1}{12}h^3 + \mathcal{O}(h^5)$$

on each step while the exact solution rotates through an angle h. Hence, after n steps where $nh = t^*$, the numerical solution underrotates by an angle $\frac{1}{12}t^*h^2 + \mathcal{O}(h^4)$—this is known as the phase error.

7.8.** The matrix

$$A(\alpha) = \begin{bmatrix} \cos\alpha & -\sin\alpha \\ \sin\alpha & \cos\alpha \end{bmatrix}$$

is known as a rotation matrix since $A\boldsymbol{u}$ rotates a general vector $\boldsymbol{u} \in \mathbb{R}^2$ counter clockwise through an angle α. Show that (a) $\det A(\alpha) = 1$, (b) $A(\alpha)A(\beta) = A(\alpha + \beta)$ and (c) $A(\alpha)^{-1} = A(-\alpha)$.

Show also that the trapezoidal rule applied to the system in Example 7.6 may be written in matrix-vector form as

$$A(-\alpha)\boldsymbol{u}_{n+1} = A(\alpha)\boldsymbol{u}_n,$$

where $\alpha = \tan^{-1}(\frac{1}{2}h)$. Deduce that the numerical solution rotates through an angle 2α in each time step in agreement with the previous exercise.

7.9.** Consider the complex IVP $x'(t) = ix(t)$ with $x(0) = 1$ that was introduced in Section 7.3. Show that the roots of the stability polynomial ($\lambda = i$, $\widehat{h} = ih$) for the mid-point rule (Example 6.12) satisfy $|r_\pm| = 1$ for $h \leq 1$ and that $|r| > 1$ for larger values of h.

7.10.*** Show that the behaviour for Simpson's rule (see Table 4.2) applied to the complex IVP $x'(t) = ix(t)$, $x(0) = 1$ is similar to that of the mid-point rule in the previous exercise. What is the largest value of h for which the roots of its stability polynomial satisfy $|r| = 1$?

8

Linear Multistep Methods—V: Solving Implicit Methods

8.1 Introduction

The discussion of absolute stability in previous chapters shows that it can be advantageous to use an implicit LMM—usually when the step size in an explicit method has to be chosen on grounds of stability rather than accuracy. One then has to compute the numerical solution at each step by solving a nonlinear system of algebraic equations. For example, when a k-step LMM is used to solve the IVP

$$\left. \begin{array}{l} \boldsymbol{x}'(t) = \boldsymbol{f}(t, \boldsymbol{x}(t)), \quad t > t_0 \\ \boldsymbol{x}(t_0) = \boldsymbol{\eta} \end{array} \right\}, \tag{8.1}$$

$\boldsymbol{x}_{n+k}$ is computed from

$$\boldsymbol{x}_{n+k} + \alpha_{k-1}\boldsymbol{x}_{n+k-1} + \cdots + \alpha_0\boldsymbol{x}_n = h\big(\beta_k \boldsymbol{f}_{n+k} + \beta_{k-1}\boldsymbol{f}_{n+k-1} \cdots + \beta_0\boldsymbol{f}_n\big), \tag{8.2}$$

with $\boldsymbol{f}_{n+k} = \boldsymbol{f}(t_{n+k}, \boldsymbol{x}_{n+k})$. Defining

$$\boldsymbol{g}_n = h\big(\beta_{k-1}\boldsymbol{f}_{n+k-1} \cdots + \beta_0\boldsymbol{f}_n\big) - \alpha_{k-1}\boldsymbol{x}_{n+k-1} - \cdots - \alpha_0\boldsymbol{x}_n,$$

which is entirely comprised of known quantities, then $\boldsymbol{x}_{n+k}$ is the solution $\boldsymbol{u}$ of the nonlinear equation

$$\boldsymbol{u} = h\beta_k \boldsymbol{f}(t_{n+k}, \boldsymbol{u}) + \boldsymbol{g}_n. \tag{8.3}$$

D.F. Griffiths, D.J. Higham, *Numerical Methods for Ordinary Differential Equations*,
Springer Undergraduate Mathematics Series, DOI 10.1007/978-0-85729-148-6_8,
© Springer-Verlag London Limited 2010

This equation always has a solution when $h = 0$ ($\boldsymbol{u} = \boldsymbol{g}_n$) and we shall assume that this continues to be so when h is sufficiently small. Since this is a nonlinear equation, it may have zero, one or more solutions[1]—in the latter case it makes sense to choose the solution closest to $\boldsymbol{x}_{n+k-1}$, the value at the previous step.

The approach taken to solve Equation (8.3) depends on the nature of the problem: if stiffness (see Section 7.2) is not an issue then we shall use either a fixed-point iteration or pairs of LMMs called predictor-corrector pairs, otherwise the Newton–Raphson method will be used. These are described in the following sections.

8.2 Fixed-Point Iteration

fixed-point iteration, also known as Picard iteration or the method of successive substitutions, involves making an initial guess, $\boldsymbol{u}^{[0]}$ say, and substituting this into the right-hand side of (8.3), thereby producing the next approximation to the root. Generally, the next approximation, $\boldsymbol{u}^{[\ell+1]}$, is computed from $\boldsymbol{u}^{[\ell]}$ using

$$\boldsymbol{u}^{[\ell+1]} = h\beta_k \boldsymbol{f}(t_{n+k}, \boldsymbol{u}^{[\ell]}) + \boldsymbol{g}_n, \quad \ell = 0, 1, 2, \ldots. \tag{8.4}$$

There are a number of immediate issues:

1. Choice of initial guess $\boldsymbol{u}^{[0]}$. Typically, the closer we can choose this to the (unknown) value of $\boldsymbol{x}_{n+k}$ the fewer iterations will be required to obtain an accurate approximation. An obvious choice is to use the solution $\boldsymbol{x}_{n+k-1}$ from the previous time step. In the next section we describe an improvement on this by making use of a "predictor".

2. Does the sequence $\boldsymbol{u}^{[\ell]}$ converge? To analyse this, we suppose that $\boldsymbol{u}^{[\ell]} = \boldsymbol{x}_{n+k} + \boldsymbol{E}^{[\ell]}$ and then, using the vector form of the Taylor expansion (C.3) in Appendix C, we find

$$\boldsymbol{f}(t_{n+k}, \boldsymbol{u}^{[\ell]}) = \boldsymbol{f}(t_{n+k}, \boldsymbol{x}_{n+k} + \boldsymbol{E}^{[\ell]})$$
$$\approx \boldsymbol{f}(t_{n+k}, \boldsymbol{x}_{n+k}) + \frac{\partial \boldsymbol{f}}{\partial \boldsymbol{x}}(t_{n+k}, \boldsymbol{x}_{n+k})\boldsymbol{E}^{[\ell]}. \tag{8.5}$$

Substituting this into the right-hand side of (8.4) and subtracting Equation (8.3) from the result gives

$$\boldsymbol{E}^{[\ell+1]} \approx h\beta_k B \boldsymbol{E}^{[\ell]}, \tag{8.6}$$

where we have used

$$B = \frac{\partial \boldsymbol{f}}{\partial \boldsymbol{x}}(t_{n+k}, \boldsymbol{x}_{n+k})$$

---

[1]This issue was addressed for some scalar problems in Exercises 4.2–4.4.

to denote the Jacobian of f at the point (t_{n+k}, x_{n+k}). We can get some insight by observing that if λ_B is an eigenvalue of B with corresponding eigenvector v, then choosing $E^{[0]} = v$ and assuming equality in (8.6) we have

$$E^{[\ell]} = \left(h\beta_k \lambda_B\right)^{\ell} v.$$

It follows that $E^{[\ell]}$ cannot tend to zero as $\ell \to \infty$ unless

$$h|\beta_k \lambda_B| < 1$$

for each eigenvalue λ_B of B (see, for example, Kelley [41, Theorem 1.3.2 and Chapter 4] for a more complete analysis). This condition tells us, in principle, how small h needs to be in order for the fixed-point iteration (8.4) to converge. In practice, it is too expensive to calculate the Jacobian and its eigenvalues, but what we can glean from this condition is that there is a restriction on h not dissimilar to that required for absolute stability. It is for this reason that fixed-point iteration is not suitable for stiff problems.

3. Termination of the iteration. This is a rather delicate issue and we direct interested readers to the book of Dahlquist and Björk [17, Chapter 6] for a detailed discussion in the scalar case. A rather crude criterion is to terminate the iteration when the difference between successive iterations is sufficiently small:

$$\|E^{[\ell+1]} - E^{[\ell]}\| \le \varepsilon,$$

for some small positive number ε, though this can be give a misleading impression of closeness to the solution when the iteration is slowly convergent (see Exercise 8.16).

The above discussion tends to mitigate against the use of fixed-point iterations, and an alternative is described in the next section.

Example 8.1

Use the backward Euler method with step length $h = 0.1$ to calculate an approximate solution to the IVP $x'(t) = 2x(t)(1 - x(t))$, $x(0) = 1/5$ at $t = h$.

The backward Euler method applied to this IVP leads to

$$x_{n+1} = x_n + 2hx_{n+1}(1 - x_{n+1}), \qquad n = 0, 1, 2, \ldots, \tag{8.7}$$

with $t_0 = 0$ and $x_0 = 1/5$. Then x_1 is the solution of the nonlinear equation

$$u = 0.2 + 0.2u(1 - u). \tag{8.8}$$

With a starting guess, $u^{[0]} = 0.2$, successive approximations are calculated from the iteration

$$u^{[\ell+1]} = 0.2 + 0.2u^{[\ell]}(1 - u^{[\ell]}), \quad \ell = 0, 1, 2, \ldots$$

and shown in Table 8.1.

ℓ	0	1	2	3	4
$u^{[\ell]}$	0.2	0.232	0.2356	0.2360	0.2360
$u^{[\ell+1]} - u^{[\ell]}$	0.032	0.0036	0.0004	0.0000	

Table 8.1 Results showing the convergence $u^{[\ell]} \to x_1$ for Example 8.1

The iteration clearly converges and $u^{[3]}$ and $u^{[4]}$ agree to four decimal places, so we can take $x_1 \approx u^{[4]} = 0.2360$. It should be noted that each iterate has about one more digit of accuracy than its predecessor, the explanation for this is left to Exercise 8.3. In this example (8.7) is a quadratic equation that may be solved directly. Its roots are $0.236\,07\ldots$ (agreeing with the value calculated above) and -4.2361, which is clearly spurious in this case.

Note that

$$u^{[\ell+1]} - u^{[\ell]} = 0.2 + 0.2u^{[\ell]}(1 - u^{[\ell]}) - u^{[\ell]}. \tag{8.9}$$

The right-hand side is the residual when $y = u^{[\ell]}$ is substituted into Equation (8.8) and is a measure of how well $u^{[\ell]}$ satisfies the equation. □

8.3 Predictor-Corrector Methods

There are many variants of predictor-corrector methods. We will restrict ourselves to describing the simplest (and perhaps the most commonly used) version. They are designed to address two of the three main issues raised in the previous section. They do this by using a pair of LMMs: one explicit and one implicit. In our examples they will both be of the same order of accuracy, p, say. The forward and backward Euler methods are a possible first-order pair and the combination of AB(2) and trapezoidal rule is a popular second-order pair. More generally, pairs of Adams–Bashforth and Adams–Moulton methods (see Section 5.2) of the same order of accuracy can be used.

We suppose that the explicit method is given by

$$x_{n+k} + \alpha_{k-1}^* x_{n+k-1} + \cdots + \alpha_0^* x_n = h\big(\beta_{k-1}^* f_{n+k-1} \cdots + \beta_0^* f_n\big), \tag{8.10}$$

with error constant C_{p+1}^*, and the implicit method by (8.2) with error constant C_{p+1}. Since we have supposed that the two methods have the same order, it usually happens that $\alpha_0 = \beta_0 = 0$ in (8.2) so that, strictly speaking, the implicit method has step number $k - 1$.

The computation of x_{n+k} proceeds by

1. using the explicit LMM (8.10) to determine a "predicted" value which we denote by $x_{n+k}^{[0]}$,

2. evaluating the right-hand side of the ODE with this value: $f_{n+k}^{[0]} = f(t_{n+k}, x_{n+k}^{[0]})$;

3. calculating the value of x_{n+k} by replacing f_{n+k} by $f_{n+k}^{[0]}$ on the right of (8.2);

4. evaluating the right-hand side of the ODE with this value: $f_{n+k} = f(t_{n+k}, x_{n+k})$.

The four steps of this algorithm are known by the acronym PECE for predict, evaluate, correct and evaluate.

One of the by-products of using predictor and corrector formulae of the same order is that it can be shown that (see Exercise 8.18 and Lambert's book [44, Chapter 4])

$$\frac{C_{p+1}}{C_{p+1}^* - C_{p+1}}\left(x_{n+k} - x_{n+k}^{[0]}\right) \tag{8.11}$$

provides an estimate of the leading term in the LTE of the corrector—this is known as Milne's device. If the computation proceeds with

$$x_{n+k} + \frac{C_{p+1}}{C_{p+1}^* - C_{p+1}}\left(x_{n+k} - x_{n+k}^{[0]}\right)$$

instead of x_{n+k} then the process is called *local extrapolation*. This updated value will be accurate of order $\mathcal{O}(h^{p+1})$ and be expected to improve the accuracy of the numerical solution. In these cases the evaluation of f_{n+k} should be carried out after this update rather than at stage 4 listed above.

Estimates of the LTE, such as that given by (8.11), can also prove to be useful in methods that vary the step length h from one step to the next (see Exercise 11.13).

Example 8.2

Use the forward and backward Euler methods as a predictor-corrector pair to calculate x_1 for the IVP of Example 8.1 with $h = 0.1$. Use Milne's device to estimate the LTE at the end of this step and, hence, find a higher order approximation to x_1.

With $f(t, x) = 2x(1 - x)$, the steps of the PECE method are, with $x_0 = 0.2$, $f_0 = 2x_0(1 - x_0) = 0.32$ and $h = 0.1$:

$$
\begin{aligned}
&\text{P: } x_1^{[0]} = x_0 + 0.1 f_0 && = 0.232, \\
&\text{E: } f_1^{[0]} = f(t_1, x_1^{[0]}) && = 0.356\,35, \\
&\text{C: } x_1 = x_0 + 0.1 f_1^{[0]} && = 0.235\,64, \\
&\text{E: } f_1 = f(t_1, x_1) && = 0.360\,23.
\end{aligned}
$$

The forward and backward Euler methods have order $p = 1$ and error constants $C_2^* = 1/2$ and $C_2 = -1/2$ respectively. The estimate (8.11) of the LTE gives, in this case,

$$-\tfrac{1}{2}\left(x_1 - x_1^{[0]}\right) = -0.001\,82,$$

When this is added to the above value of x_1, we obtain the second order accurate value 0.233 72. It can be shown that the exact solution of the IVP at $t = t_1$ is $x(t_1) = 0.233\,92$ and the updated approximation is seen to be accurate to three significant figures.   □

The absolute stability properties of predictor-corrector pairs can be deduced as described in Exercises 8.7 and 8.15. The regions of absolute stability are generally much closer to those of the predictor than to the corrector. They are explicit methods (since they do not involve the solution of equations to determine x_{n+k}) and so cannot be A-stable.

8.4 The Newton–Raphson Method

To apply the Newton–Raphson method, we write (8.3) as $F(u) = 0$, where the function F is defined by

$$F(u) = u - h\beta_k f(t_{n+k}, u) - g_n, \qquad (8.12)$$

whose solution is $x_{n+k} = u$. Suppose that we have an approximation $u^{[\ell]}$ to x_{n+k} and we define $E^{[\ell]} = u^{[\ell]} - x_{n+k}$ as in Section 8.2. Then, since $F(x_{n+k}) = 0$, we have by Taylor expansion (see Appendix C)

$$
\begin{aligned}
0 = F(x_{n+k}) \\
= F(u^{[\ell]} - E^{[\ell]}) \\
\approx F(u^{[\ell]}) - \frac{\partial F}{\partial x}(u^{[\ell]}) E^{[\ell]}.
\end{aligned}
$$

Supposing that the Jacobian is nonsingular,[2] the solution $\widehat{\boldsymbol{E}}^{[\ell]}$ of the (linear) system of algebraic equations

$$\frac{\partial \boldsymbol{F}}{\partial \boldsymbol{x}}(\boldsymbol{u}^{[\ell]})\widehat{\boldsymbol{E}}^{[\ell]} = \boldsymbol{F}(\boldsymbol{u}^{[\ell]}) \tag{8.13}$$

is expected to be close to $\boldsymbol{E}^{[\ell]}(= \boldsymbol{u}^{[\ell]} - \boldsymbol{x}_{n+k})$ and an improved approximation for the solution is then given by

$$\boldsymbol{u}^{[\ell+1]} = \boldsymbol{u}^{[\ell]} - \widehat{\boldsymbol{E}}^{[\ell]}. \tag{8.14}$$

The Newton–Raphson method converges very rapidly (under reasonable assumptions) provided that the initial guess is sufficiently close to $\boldsymbol{x}_{n+k}$—in fact, convergence is quadratic, in the sense that the magnitude $\|\boldsymbol{E}^{[\ell+1]}\|$ of the distance from the new approximation to the exact solution is proportional to $\|\boldsymbol{E}^{[\ell]}\|^2$ (see, for example, Kelley [41, Chapter 5]). In practical terms this is interpreted as saying that if the ℓth approximation is correct to d digits, say, then the $(\ell + 1)$th is accurate to $2d$ digits—the number of correct digits doubles at each iteration. The cost of this rapid convergence is, of course, the need to compute the Jacobian matrix and then solve the linear system of equations (8.13) at each iteration.

Unlike the fixed-point iteration method, if the nonlinear equations are solved accurately at each step, then the absolute stability characteristics of the implicit LMM are maintained.

Example 8.3

Use the Newton–Raphson method to find an approximation to the IVP

$$\begin{aligned}
x'(t) &= -2y(t)^3, & x(0) &= 1, \\
y'(t) &= 2x(t) - y(t)^3, & y(0) &= 1,
\end{aligned} \tag{8.15}$$

at $t = h$ with the backward Euler method and a step length $h = 0.1$.

Using $\boldsymbol{u} = [u, v]^{\mathrm{T}}$ to represent the solution $\boldsymbol{x}_{n+1}$, the backward Euler method leads to the equations

$$\begin{aligned}
u &= x_n - 2hv^3, \\
v &= y_n + h(2u - v^3).
\end{aligned} \tag{8.16}$$

At $n = 0$, we have to solve $\boldsymbol{F}(\boldsymbol{u}) = \boldsymbol{0}$, where

$$\boldsymbol{F}(\boldsymbol{u}) = \begin{bmatrix} u - 1 + 2hv^3 \\ v - 1 - h(2u - v^3) \end{bmatrix}.$$

---

[2]The Jacobian of $\boldsymbol{F}$ is related to that of $\boldsymbol{f}$ by $\frac{\partial \boldsymbol{F}}{\partial \boldsymbol{x}}(\boldsymbol{u}) = \boldsymbol{I} - h\beta_k \frac{\partial \boldsymbol{f}}{\partial \boldsymbol{x}}(t_{n+k}, \boldsymbol{u})$. This will be nonsingular if h is sufficiently small.

ℓ	0	1	2	3
$\boldsymbol{u}^{[\ell]}$	1.00	0.774\,647\,887	0.773\,901\,924	0.773\,901\,807
	1.00	1.042\,253\,521	1.041\,731\,347	1.041\,731\,265
$\widehat{\boldsymbol{E}}^{[\ell]}$	2.2535×10^{-1}	7.4596×10^{-4}	1.1711×10^{-7}	2.9131×10^{-15}
	-4.2254×10^{-2}	5.2217×10^{-4}	8.1975×10^{-8}	1.9628×10^{-15}
$\boldsymbol{E}^{[\ell]}$	2.2610×10^{-1}	7.4608×10^{-4}	1.1711×10^{-7}	2.8866×10^{-15}
	-4.1731×10^{-2}	5.2226×10^{-4}	8.1975×10^{-8}	1.9984×10^{-15}

Table 8.2   Results illustrating convergence of the Newton–Raphson iteration to determine $\boldsymbol{x}_1$ in Example 8.15

The Jacobian of $\boldsymbol{F}(\boldsymbol{u})$ is given by

$$\frac{\partial \boldsymbol{F}}{\partial \boldsymbol{x}}(\boldsymbol{u}) = \begin{bmatrix} 1 & 6hv^2 \\ -2h & 1 + 3hv^2 \end{bmatrix}$$

so a typical iteration of the Newton–Raphson method involves solving

$$\begin{bmatrix} 1 & 6h(v^{[\ell]})^2 \\ -2h & 1 + 3h(v^{[\ell]})^2 \end{bmatrix} \widehat{\boldsymbol{E}}^{[\ell]} = \boldsymbol{F}(\boldsymbol{u}^{[\ell]})$$

and then setting $\boldsymbol{u}^{[\ell+1]} = \boldsymbol{u}^{[\ell]} - \widehat{\boldsymbol{E}}^{[\ell]}$, for $\ell = 0, 1, 2, \ldots$ with $\boldsymbol{u}^{[0]} = [1, 1]^{\mathrm{T}}$. The results are summarized in Table 8.2. Not only have we shown more iterations than necessary, we have also included more digits in each entry than the accuracy of the method warrants. This has been done to fully illustrate how rapidly convergence occurs and to show that convergence is indeed quadratic (see Exercise 8.13). The last two rows of the table have been computed by preforming two further iterations and regarding the result as being the exact solution to (8.16). It is also seen from Table 8.2 that $\widehat{\boldsymbol{E}}^{[\ell]} \approx \boldsymbol{E}^{[\ell]}$—this is a consequence of the rapid convergence. Since the underlying method is only first-order accurate and the grid size $h = 0.1$ is quite large, the first iterate $\boldsymbol{u}^{[1]}$ is already sufficiently accurate. $\square$

8.5 Postscript

The efficient treatment of implicit methods is an essential requirement for dealing with stiff systems of ODEs. We have given an introduction to the basic ideas, but further refinements are common. For example, using an explicit LMM as a predictor will generally provide a better starting value for fixed-point iterations. Also, at the end of Section 8.2 it was shown in (8.9) that $u^{[\ell+1]} - u^{[\ell]}$ was equal to the residual when $u = u^{[\ell]}$ was substituted into the implicit LMM.

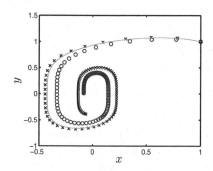

Fig. 8.1 Numerical solutions for Example 8.3 integrated over $0 \leq t \leq 10$. The circles show the solution with backward Euler and the crosses the solution with local extrapolation, both with $h = 0.1$. The solid curve shows an accurate solution computed with an RK method

Recalling that the LTE is obtained when the exact solution is substituted into the same equation, it makes sense for the iteration process to be terminated when $u^{[\ell+1]} - u^{[\ell]}$ is of the same magnitude as the LTE. The predicted value can then be used, via Milne's device, to estimate the LTE.

The idea of using an explicit predictor with Milne's device can also be used with the Newton–Raphson method. In Example 8.3, the use of Euler's method to predict a value of $\boldsymbol{x}_1$ would give

$$\boldsymbol{x}_1^{[0]} = \boldsymbol{x}_0 + h\boldsymbol{f}(t_0, \boldsymbol{x}_0) = \begin{bmatrix} 0.8 \\ 1.1 \end{bmatrix}.$$

The earlier use of the Newton–Raphson iteration to solve the backward Euler equations gave $\boldsymbol{x}_1 = [0.7739, 1.0417]^{\mathrm{T}}$ (see Table 8.2) and these two approximations can be combined in Milne's device (Equation (8.11) with $p = 1$, $C_2^* = \frac{1}{2}$, and $C_2 = -\frac{1}{2}$) to give the LTE estimate

$$-\tfrac{1}{2}(\boldsymbol{x}_1 - \boldsymbol{x}_1^{[0]}) = \begin{bmatrix} 0.013 \\ 0.029 \end{bmatrix}.$$

For local extrapolation this is added to the backward Euler solution to give the updated solution $\boldsymbol{x}_1 = [0.7752, 1.0707]^{\mathrm{T}}$ which lies much closer to the exact solution $\boldsymbol{x}(t_1) = [0.7757, 1.0661]^{\mathrm{T}}$ (to four decimal places).

The system is now integrated over the time interval $0 \leq t \leq 10$ and the resulting phase plane solutions are shown in Figure 8.1. The numerical solution using the backward Euler method is shown by circles (∘), the solution with local extrapolation by crosses (×), and an accurate numerical solution (computed with the MATLAB command ode45) as a solid curve. The extrapolated solution, which was obtained with little additional computational cost, is significantly more accurate than the backward Euler solution.   □

EXERCISES

8.1.** Solve the quadratic equation (8.7) for x_{n+1} in terms of x_n and h. Discuss the behaviour of these solutions as $h \to 0$.

8.2.* Extend the calculation in Example 8.2 to determine x_2. What is the improved value given by Milne's device?

8.3.* Calculate $E^{[\ell]}(= u^{[\ell]} - x_1)$ for the data in Table 8.1 for $\ell = 0 : 3$ on the basis that $x_1 = 0.23607$. Show that these values have the property that $E^{[\ell+1]}/E^{[\ell]}$ is approximately constant and the value of this constant is approximately hB, where B is an appropriate Jacobian evaluated at $x = x_1$, thus confirming the approximation (8.6). Deduce that successive iterates gain approximately one extra digit of accuracy.

8.4.* Apply the backward Euler method with $h = 0.1$ to solve the IVP described in Example 8.3. Calculate the first two fixed-point iterations in the determination of $\boldsymbol{x}_1$ from a starting guess of $\boldsymbol{u}^{[0]} = \boldsymbol{x}_0$.

8.5.* For the system of ODEs in Example 8.3, show that

$$\frac{d}{dt}\left(x(t)^2 + \tfrac{1}{2}y(t)^4\right) = -2y(t)^6.$$

The quantity $V(t) \equiv x(t)^2 + \tfrac{1}{2}y(t)^4$ is an example of a Lyapunov function. Its time derivative is negative except when $y(t) = 0$, in which case, $x'(t) = 0$ and $y'(t) = 2x(t)$—so the motion is vertical and counterclockwise in the phase plane. $V(t)$ is therefore a decreasing function of t, from which it can be concluded that, in the long term, the solution spirals to the origin.

8.6.** Apply the forward and backward Euler methods as a predictor-corrector pair to approximate $\boldsymbol{x}_1$ for the IVP described in Example 8.3 with $h = 0.1$. Compare your answers with those given in Table 8.2.

Use Milne's device to produce a revised estimate of the solution at $t = t_1$.

8.7.** Apply the forward and backward Euler methods as a predictor-corrector pair to solve the model equation $x'(t) = \lambda x(t)$ and show that

$$x_{n+1} = (1 + \widehat{h} + \widehat{h}^2)x_n,$$

where $\widehat{h} = h\lambda$. Hence, show that the interval of absolute stability of this method is $\widehat{h} \in (-1, 0)$. The region of absolute stability is shown in Figure 8.2 (left).

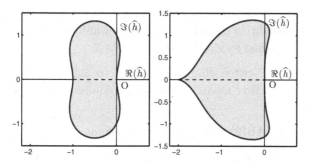

Fig. 8.2 The region of absolute stability for the forward/backward Euler PECE method for Exercise 8.7 (left) and the AB(2)/trapezoidal pair for Exercise 8.15 (right)

8.8.* (Based on Stuart and Humphries [65, page 270].) Applying the backward Euler method to the ODE $x'(t) = -x^3(t)$ requires us to find a root of the cubic equation $g(u) = u - x_n + hu^3$. Show that, for any fixed $h > 0$ and x_n,

(i) $g(u) \to -\infty$ as $u \to -\infty$;

(ii) $g(u) \to \infty$ as $u \to \infty$;

(iii) $g(u)$ is monotonically increasing.

Deduce that there is always a unique solution for the backward Euler method in this case.

This example is studied further in Exercise 14.12.

8.9.** The scalar function f is said to satisfy a *one-sided Lipschitz condition* if there exists a constant γ such that, for all u, v,

$$(u - v)(f(u) - f(v)) \le \gamma(u - v)^2.$$

Consider applying backward Euler to the ODE $x'(t) = f(x)$. Show that when f satisfies a one-sided Lipschitz condition there is always a unique solution if $h\gamma < 1$. In particular, this implies that there is a unique solution for any step size when $\gamma \le 0$. (Hint: if there are two distinct solutions, u and v, then $u = x_n + hf(u)$ and $v = x_n + hf(v)$. Subtract one equation from the other, multiply both sides by $u - v$ and apply the one-sided Lipschitz condition.) These ideas can be generalized for systems of ODEs; see, for example, Stuart and Humphries [65].

8.10.** Following on from Exercises 8.8 and 8.9, show that $f(u) = -u^3$ satisfies a one-sided Lipschitz condition with $\gamma = 0$. Also show that

$f(u) = u - u^3$ satisfies a one-sided Lipschitz condition and, hence, find a condition on h that guarantees a unique solution for backward Euler applied to $x'(t) = x(t) - x^3(t)$.

8.11.* Use the Newton–Raphson method to solve Equation (8.8) with $u^{[0]} = 0.2$. How many iterations are required so that $|u^{[\ell+1]} - u^{[\ell]}| < 0.001$?

8.12.** Use the AB(2) and trapezoidal methods

$$x_{n+2} = x_{n+1} + \tfrac{1}{2}h(3f_{n+1} - f_n), \qquad C_3^* = \tfrac{5}{12},$$
$$x_{n+2} = x_{n+1} + \tfrac{1}{2}h(f_{n+2} + f_{n+1}), \qquad C_3 = -\tfrac{1}{12},$$

as a predictor-corrector pair to calculate x_2 for the IVP of Example 8.1 with $h = 0.1$ and the extra starting value $x_1 = 0.233\,922$. Use the error constants shown above with Milne's device to estimate the LTE at the end of this step and incorporate this to find a higher order approximation to x_2. How do these two solutions compare with the exact solution $x(t_2) = 0.271\,645$ (to six decimal places).

This process is generalized in Exercise 11.13 to variable step sizes.

8.13.* Suppose that $\boldsymbol{E}^{[\ell]} = [p^{[\ell]}, q^{[\ell]}]^{\mathrm{T}}$. Use the data provided in the last two rows of Table 8.2 to show that

$$\frac{(p^{[\ell]})^2}{p^{[\ell+1]}} \approx 4.75$$

for $\ell = 1, 2$, thus confirming quadratic convergence. Show that a similar result holds for the second component $q^{[\ell]}$.

8.14.** Write down the equations that have to be solved when the trapezoidal rule is used to compute an approximation to $\boldsymbol{x}(t_1)$ for the IVP described in Example 8.3 with $h = 0.1$. Carry out two iterations when the Newton–Raphson method is used to solve these equations starting from the initial guess $\boldsymbol{x}(0) = [1, 1]^{\mathrm{T}}$. How does the numerical solution at this stage compare with the values obtained in Exercise 8.6 from the backward Euler method with and without local extrapolation?

8.15.*** Show that the PECE pair comprising AB(2) and trapezoidal methods of Exercise 8.12 applied to ODE $x'(t) = \lambda x(t)$ leads to

$$x_{n+2} = (1 + \widehat{h} + \tfrac{3}{4}\widehat{h}^2)x_{n+1} - \tfrac{1}{4}\widehat{h}^2 x_n.$$

Hence, show that the the interval of absolute stability of this method is $\widehat{h} \in (-2, 0)$. The region of absolute stability is shown in Figure 8.2 (right).

8.16.* Suppose that C is a positive constant, $-1 < r < 1$, and v a unit vector. The sequence $E^{[\ell]} = Cr^\ell v$ converges geometrically to zero (behaviour typical of fixed-point iterations). Show that

$$\|E^{[\ell+1]} - E^{[\ell]}\| = (1 - r)\|E^{[\ell]}\|$$

and so the distance to the limit is given by

$$\|E^{[\ell]}\| = \frac{\varepsilon}{1 - r}$$

when $\|E^{[\ell+1]} - E^{[\ell]}\| = \varepsilon$. Thus, when the iteration is slowly convergent (r is close to, but smaller than, unity), the distance from the limit may be considerably further than the distance between successive iterates. Thus, to ensure that $\|E^{[\ell]}\| < \delta$, iterations should continue until $\|E^{[\ell+1]} - E^{[\ell]}\| < (1 - r)\delta$.

8.17.* Using the values of $E^{[\ell]}$ and $r = hB$ calculated in Exercise 8.3, verify that the data in Table 8.1 satisfy (approximately) the result

$$|E^{[\ell]}| = \frac{|E^{[\ell+1]} - E^{[\ell]}|}{1 - r}$$

that was derived in the previous exercise.

8.18.** It was shown in Section 5.4 that, for linear ODEs, the leading term in the LTE for both explicit and implicit two-step LMMs was equal to the difference between the computed approximation $\tilde{x}$ and the exact solution $x(t_{n+2})$, under the localizing assumption that all back values are exact. This result holds more generally and, updating the notation to match the predictor-corrector setting, we have, under the localizing assumption,

$$\text{P:} \quad x(t_{n+k}) - x^{[0]}_{n+k} = T^*_{n+k}$$
$$= h^{p+1}C^*_{p+1}\frac{d^{p+1}x}{dt^{p+1}}(t_{n+k})$$
$$\text{C:} \quad x(t_{n+k}) - x_{n+k} = T_{n+k} + \mathcal{O}(h^{p+2})$$
$$= h^{p+1}C_{p+1}\frac{d^{p+1}x}{dt^{p+1}}(t_{n+k}) + \mathcal{O}(h^{p+2}).$$

By ignoring higher order terms, express $h^{p+1}x^{(p+1)}(t_{n+k})$ in terms of $x^{[0]}_{n+k}$ and x_{n+k}. Hence, show that the leading term in T_{n+k} is given by (8.11).

9
Runge–Kutta Method—I:
Order Conditions

9.1 Introductory Examples

Runge–Kutta (RK) methods are one-step methods composed of a number of stages. A weighted average of the slopes (f) of the solution computed at nearby points is used to determine the solution at $t = t_{n+1}$ from that at $t = t_n$. Euler's method is the simplest such method and involves just one stage.

As in earlier chapters, we will develop methods for solving the IVP

$$\left. \begin{array}{c} x'(t) = f(t, x(t)), \quad t > t_0 \\ x(t_0) = \eta \end{array} \right\}. \tag{9.1}$$

We give two examples of RK methods before going on to describe the general method.

Example 9.1 (An Explicit RK Method)

Use the two-stage RK method (known as the "modified Euler method") given by

$$\begin{aligned} k_1 &= f(t_n, x_n), \\ k_2 &= f(t_n + \tfrac{1}{2}h, x_n + \tfrac{1}{2}hk_1), \\ x_{n+1} &= x_n + hk_2, \\ t_{n+1} &= t_n + h \end{aligned}$$

D.F. Griffiths, D.J. Higham, *Numerical Methods for Ordinary Differential Equations*,
Springer Undergraduate Mathematics Series, DOI 10.1007/978-0-85729-148-6_9,
© Springer-Verlag London Limited 2010

to calculate approximate solutions to the IVP (see Example 2.1)

$$x'(t) = (1 - 2t)x(t), \quad t > 0,$$
$$x(0) = 1 \tag{9.2}$$

at $t = h$ and $t = 2h$ using $h = 0.2$.

We find that

$$k_1 = f(t_n, x_n) = (1 - 2t_n)x_n,$$
$$k_2 = f(t_n + \tfrac{1}{2}h, x_n + \tfrac{1}{2}hk_1)$$
$$= (1 - 2t_n - h)(x_n + \tfrac{1}{2}hk_1).$$

So, with $h = 0.2$:

$$n = 0: \quad k_1 = (1 - 2t_0)x_0 = 1,$$
$$k_2 = (1 - 2t_0 - h)(x_0 + \tfrac{1}{2}hk_1) = 0.8(1.1) = 0.88,$$
$$x_1 = x_0 + hk_2 = 1.176,$$
$$t_1 = t_0 + 0.2 = 0.2.$$
$$n = 1: \quad k_1 = (1 - 2t_1)x_1 = 0.6 \times 1.176 = 0.7056,$$
$$k_2 = (1 - 2t_1 - h)(x_1 + \tfrac{1}{2}hk_1) = 0.498\,62,$$
$$x_2 = x_1 + hk_2 = 1.176 + 0.2 \times 0.498\,62 = 1.2757,$$
$$t_2 = t_1 + 0.2 = 0.4.$$

This is an example of an *explicit* RK method: each k can be calculated without having to solve any equations. When the calculations are extended to $t = 1.2$ and repeated with $h = 0.1$ we find the GEs shown in the final column of Table 9.1. The GEs for this method are comparable with those from the second-order methods of earlier chapters.   □

h	TS(2)	Trap.	ABE	ABT	RK(2)
0.2	5.4	−2.8	−3.6	17.6	3.5
0.1	1.4	−0.71	−0.66	4.0	0.67
Ratio	3.90	4.00	5.49	4.40	5.24

Table 9.1 Global errors (multiplied by 10^3) at $t = 1.2$ for the RK(2) method of Example 9.1 alongside the GEs for the second-order methods used in Table 4.1

Example 9.2 (An Implicit RK Method)

For the implicit RK method defined by

$$k_1 = f(t_n + (\tfrac{1}{2} - \gamma)h, x_n + \tfrac{1}{4}hk_1 + (\tfrac{1}{4} - \gamma)hk_2),$$
$$k_2 = f(t_n + (\tfrac{1}{2} + \gamma)h, x_n + (\tfrac{1}{4} + \gamma)hk_1 + \tfrac{1}{4}hk_2),$$
$$x_{n+1} = x_n + \tfrac{1}{2}h(k_1 + k_2),$$

where $\gamma = \sqrt{3}/6$, determine x_{n+1} in terms of x_n when applied to the IVP $x'(t) = \lambda x(t)$ with $x(0) = 1$.

With $f(t, x) = \lambda x$, we obtain

$$k_1 = \lambda(x_n + \tfrac{1}{4}hk_1 + (\tfrac{1}{4} - \gamma)hk_2),$$
$$k_2 = \lambda(x_n + (\tfrac{1}{4} + \gamma)hk_1 + \tfrac{1}{4}hk_2).$$

Writing $\widehat{h} = h\lambda$, these equations may be rearranged to read

$$\begin{bmatrix} 1 - \tfrac{1}{4}\widehat{h} & -(\tfrac{1}{4} - \gamma)\widehat{h} \\ -(\tfrac{1}{4} + \gamma)\widehat{h} & 1 - \tfrac{1}{4}\widehat{h} \end{bmatrix} \begin{bmatrix} k_1 \\ k_2 \end{bmatrix} = \begin{bmatrix} 1 \\ 1 \end{bmatrix} \lambda x_n, \tag{9.3}$$

which may be solved for k_1 and k_2. It may then be shown that (see Exercise 9.10)

$$x_{n+1} = \frac{1 + \tfrac{1}{2}\widehat{h} + \tfrac{1}{12}\widehat{h}^2}{1 - \tfrac{1}{2}\widehat{h} + \tfrac{1}{12}\widehat{h}^2} x_n. \qquad \square \tag{9.4}$$

Implicit RK methods are obviously much more complicated than both their explicit versions and corresponding LMMs. The complexity can be justified by improved stability and higher order: the method described in the previous example is A-stable and of fourth-order (see Exercises 9.10 and 10.13).

9.2 General RK Methods

The general s-stage RK method may be written in the form

$$x_{n+1} = x_n + h \sum_{i=1}^{s} b_i k_i, \tag{9.5}$$

where the $\{k_i\}$ are computed from the function f:

$$k_i = f\left(t_n + c_i h, x_n + h \sum_{j=1}^{s} a_{i,j} k_j\right), \quad \cdot\, i = 1 : s. \tag{9.6}$$

$$
\begin{array}{c|cccc}
c_1 & a_{1,1} & a_{1,2} & \cdots & a_{1,s} \\
c_2 & a_{2,1} & a_{2,2} & \cdots & a_{2,s} \\
\vdots & \vdots & \vdots & & \\
c_s & a_{s,1} & a_{s,2} & \cdots & a_{s,s} \\
\hline
 & b_1 & b_2 & \cdots & b_s
\end{array}
$$

Table 9.2 The Butcher array for a full (implicit) RK method

Exercise 9.3 shows that it is natural to impose the condition

$$
c_i = \sum_{j=1}^{s} a_{i,j}, \qquad i = 1 : s, \tag{9.7}
$$

so we will do this throughout. Thus, given a value of s, the method depends on $s^2 + s$ parameters $\{a_{i,j}, b_j\}$. These can be conveniently displayed in a tableau known as the Butcher array—see Table 9.2. For instance, the Butcher arrays for Examples 9.1 and 9.2 are shown below:

$$
\begin{array}{c|cc}
0 & 0 & 0 \\
\frac{1}{2} & \frac{1}{2} & 0 \\
\hline
 & 0 & 1
\end{array}
\qquad
\begin{array}{c|cc}
\frac{1}{2} - \gamma & \frac{1}{4} & \frac{1}{4} - \gamma \\
\frac{1}{2} + \gamma & \frac{1}{4} + \gamma & \frac{1}{4} \\
\hline
 & \frac{1}{2} & \frac{1}{2}
\end{array}
$$

In general, Equations (9.6) constitutes s nonlinear equations to determine $\{k_i\}$; once found, these values are substituted into (9.5) to determine x_{n+1}. Thus, a general RK method is *implicit*.

However, if $a_{i,j} = 0$ for all $j \geq i$ (the matrix[1] $\mathcal{A} = (a_{i,j})$ is strictly lower triangular) the tableau is shown in Table 9.3 and $k_1, k_2, \ldots, k_s$ may be computed in turn from (9.6) without the need to solve any nonlinear equations; we say that the method is *explicit*. These are the classical RK methods. We shall

$$
\begin{array}{c|cccccc}
0 & 0 & 0 & \cdots & & & 0 \\
c_2 & a_{2,1} & 0 & \cdots & & & 0 \\
c_3 & a_{3,1} & a_{3,2} & \cdots & & & \vdots \\
\vdots & \vdots & & \ddots & 0 & & 0 \\
c_s & a_{s,1} & a_{s,2} & \cdots & a_{s,s-1} & & 0 \\
\hline
 & b_1 & b_2 & \cdots & b_{s-1} & & b_s
\end{array}
$$

Table 9.3 The Butcher array for an explicit RK method

---

[1]We use $\mathcal{A}$ in calligraphic font to distinguish the coefficient matrix of an RK method from the matrix A that appears in a model linear system of ODEs: $\boldsymbol{u}'(t) = A\boldsymbol{u}(t)$.

omit the zeros above the diagonal when writing a Butcher array for an explicit method.

In the remainder of this chapter we look at the issue of choosing parameter values to obtain the highest possible order. Chapter 10 then looks at absolute stability. We will focus almost exclusively on explicit RK methods, and hence, for brevity, "RK method" will mean "explicit RK method" by default, and we will make it clear when implicit methods are being considered.

Definition 9.3 (Local Truncation Error)

The LTE, T_{n+1}, of an RK method is defined to be the difference between the exact and the numerical solution of the IVP at time $t = t_{n+1}$:

$$T_{n+1} = x(t_{n+1}) - x_{n+1},$$

under the *localizing assumption* that $x_n = x(t_n)$, i.e. that the current numerical solution x_n is exact. If $T_{n+1} = \mathcal{O}(h^{p+1})$ ($p > 0$), the method is said to be of order p.

We note that this definition of order agrees with the versions used for TS methods (Section 3.3) and LMMs (Section 5.4).

9.3 One-Stage Methods

There is no simple equivalent of the linear difference operator for RK methods and a more direct approach is required in order to calculate the LTE.

When $s = 1$ we have $x_{n+1} = x_n + hb_1k_1$ and $k_1 = f(t_n + c_1h, x_n + a_{1,1}k_1)$. However, since we are only considering explicit methods, $c_1 = a_{1,1} = 0$ and so $k_1 = f(t_n, x_n) \equiv f_n$, leading to

$$x_{n+1} = x_n + hb_1f_n.$$

This expansion is compared with that for $x(t_{n+1})$:

$$x(t_{n+1}) = x(t_n) + hx'(t_n) + \tfrac{1}{2!}h^2x''(t_n) + \mathcal{O}(h^3).$$

Differentiating the differential equation $x'(t) = f(t, x(t))$ with respect to t using the chain rule gives

$$x''(t) = f_t + x'(t)f_x = f_t + ff_x$$

and so

$$x(t + h) = x(t) + hf + \tfrac{1}{2!}h^2(f_t + ff_x) + \mathcal{O}(h^3), \tag{9.8}$$

where f and its partial derivatives are evaluated at (t, x). Applying the localizing assumption $x_n = x(t_n)$ (so that $f = f_n$, etc.) to this when $t = t_n$ we obtain

$$
\begin{aligned}
T_{n+1} &= x(t_{n+1}) - x_{n+1} \\
&= h(1 - b_1)f_n + \tfrac{1}{2!}h^2 \left(f_t + ff_x\right)\big|_{t=t_n} + \mathcal{O}(h^3).
\end{aligned}
$$

The method will, therefore, be consistent of order $p = 1$ on choosing $b_1 = 1$, leading to Euler's method, the only first-order one-stage explicit RK method.

Of course, we could have viewed this method as a one-step LMM and come to the same conclusion that $b_1 = 1$ by applying the second of the conditions for consistency: $\rho'(1) = \sigma(1)$ (Theorem 4.7).

9.4 Two-Stage Methods

With two stages, $s = 2$, the most general form is given by (bearing in mind (9.7))

$$
\left.\begin{aligned}
k_1 &= f(t_n, x_n) \\
k_2 &= f(t_n + ah, x_n + ahk_1) \\
x_{n+1} &= x_n + h(b_1 k_1 + b_2 k_2)
\end{aligned}\right\}. \tag{9.9}
$$

We shall require the Taylor expansion of k_2 as a function of h, which, in turn, requires the Taylor expansion of a function of two variables (see Appendix C):

$$
f(t + \alpha h, x + \beta h) = f(t, x) + h(\alpha f_t(t, x) + \beta f_x(t, x)) + \mathcal{O}(h^2).
$$

Hence, since $k_1 = f_n$,

$$
f(t_n + ah, x_n + ahk_1) = f_n + ah\left(f_t + ff_x\right)\big|_{t=t_n} + \mathcal{O}(h^2). \tag{9.10}
$$

Substituting this into the equation for x_{n+1}, we obtain

$$
\begin{aligned}
x_{n+1} &= x_n + h(b_1 k_1 + b_2 k_2) \\
&= x_n + hb_1 f_n + hb_2\left(f_n + ah\left(f_t + ff_x\right)\big|_{t=t_n} + \mathcal{O}(h^2)\right) \\
&= x_n + h(b_1 + b_2)f_n + ab_2 h^2 \left(f_t + ff_x\right)\big|_{t=t_n} + \mathcal{O}(h^3). \tag{9.11}
\end{aligned}
$$

From Definition 9.3 the LTE is obtained by subtracting (9.11) from (9.8) with $t = t_n$ and assuming that $x_n = x(t_n)$:

$$
T_{n+1} = h(1 - b_1 - b_2)f_n + h^2\left(\tfrac{1}{2} - ab_2\right)\left(f_t + ff_x\right)\big|_{t=t_n} + \mathcal{O}(h^3). \tag{9.12}
$$

The terms on the right of (9.12) have a complicated structure, in contrast to the equivalent expressions for LMMs. This means that there is no direct analogue

LTE		Order conditions		Order
$\mathcal{O}(h^2)$	if	$b_1 + b_2 = 1$	for any a	$p = 1$
$\mathcal{O}(h^3)$	if	$b_1 + b_2 = 1$	$ab_2 = \frac{1}{2}$	$p = 2$

Table 9.4 The order conditions for two-stage methods

here of the error constants of LMMs. The LTE is, in general, $\mathcal{O}(h)$, giving order $p = 0$; the method is *not* consistent. Setting the coefficients of successive leading terms in (9.12) to zero gives the order conditions listed in Table 9.4.

The general two-stage RK method has three free parameters and only two conditions need be imposed to obtain consistency of order 2. It would appear that the remaining free parameter might be chosen to increase the order further. However, had we carried one further term in each expansion and applied the order 2 conditions we would have found

$$T_{n+1} = h^3\left[\left(\tfrac{1}{6}-b_2a^2\right)\left(f_{tt}+2ff_{xt}+f^2f_{xx}+f_x(f_t+ff_x)\right)+\tfrac{1}{6}f_x\left(f_t+ff_x\right)\right]+\mathcal{O}(h^4).$$

It should be clear that there are no values of the parameters a, b_1, b_2 for which the method has order 3 ($T_{n+1} = \mathcal{O}(h^4)$).

Defining $b_2 = \theta$, we have the family of two-stage RK methods with Butcher array shown in Table 9.5. All methods in the family have order 2 for $\theta \neq 0$. The most popular methods from this family are the *improved Euler method* ($\theta = \frac{1}{2}$) and the *modified Euler method* ($\theta = 1$; see Example 9.1).

$$\begin{array}{c|cc} 0 & 0 & \\ a & a & 0 \\ \hline & 1-\theta & \theta \end{array} \qquad a = 1/(2\theta)$$

Table 9.5 The Butcher array for the family of second-order, two-stage RK methods

9.5 Three–Stage Methods

The most general three-stage RK method is given by

$$\begin{aligned} k_1 &= f(t_n, x_n), \\ k_2 &= f(t_n + c_2h, x_n + ha_{2,1}k_1), \\ k_3 &= f(t_n + c_3h, x_n + ha_{3,1}k_1 + ha_{3,2}k_2), \\ x_{n+1} &= x_n + h\big(b_1k_1 + b_2k_2 + b_3k_3\big), \end{aligned}$$

corresponding to the following tableau:

$$
\begin{array}{c|ccc}
0 & 0 & & \\
c_2 & a_{2,1} & 0 & \\
c_3 & a_{3,1} & a_{3,2} & 0 \\
\hline
 & b_1 & b_2 & b_3
\end{array}
$$

It follows from (9.7) that

$$a_{2,1} = c_2 \text{ and } a_{3,1} + a_{3,2} = c_3, \qquad (9.13)$$

so the method has six free parameters.

The manipulations are significantly more complicated for constructing three-stage methods, so we shall summarize the conditions for third-order accuracy without giving any details. The strategy is to expand k_1, k_2, and k_3 about the point (t_n, x_n) and to substitute these into the equation

$$x_{n+1} = x_n + h\big(b_1 k_1 + b_2 k_2 + b_3 k_3\big)$$

so as to obtain an expansion for x_{n+1} in terms of powers of h.

This expansion is compared, term by term, with the expansion of the exact solution (see (9.8) and Exercise 9.14):

$$
\begin{aligned}
x(t + h) = x(t) + hf + \tfrac{1}{2}h^2(f_t + f f_x) \\
+ \tfrac{1}{6}h^3(f_{tt} + 2f f_{xt} + f^2 f_{xx} + f_x(f_t + f f_x)) + \mathcal{O}(h^4),
\end{aligned}
$$

where f and its partial derivatives are evaluated at (t, x). This process leads to the conditions shown in Table 9.6.

There are four nonlinear equations[2] in the six unknowns $c_2, c_3, a_{3,2}, b_1, b_2, b_3$. There is no choice of these parameters that will give a method of order greater than three.

$$
\begin{aligned}
b_1 + b_2 + b_3 &= 1 &&\text{(order 1 condition)} \\
b_2 c_2 + b_3 c_3 &= \tfrac{1}{2} &&\text{(order 2 condition)} \\
\left. \begin{array}{l} b_2 c_2^2 + b_3 c_3^2 = \tfrac{1}{3} \\[4pt] c_2 a_{3,2} b_3 = \tfrac{1}{6} \end{array} \right\} &&&\text{(order 3 conditions)}
\end{aligned}
$$

Table 9.6 Order conditions for three-stage RK methods

---

[2]When $b_3 = 0$ the first two equations give the order 2 conditions for two-stage RK methods.

Two commonly used methods from this family are Heun's and Kutta's third order rules; see Table 9.7.

0	0		
$\frac{1}{3}$	$\frac{1}{3}$	0	
$\frac{2}{3}$	0	$\frac{2}{3}$	0
	$\frac{1}{4}$	0	$\frac{3}{4}$

0	0		
$\frac{1}{2}$	$\frac{1}{2}$	0	
1	-1	2	0
	$\frac{1}{6}$	$\frac{4}{3}$	$\frac{1}{6}$

Table 9.7 Heun's third order rule & Kutta's third order rule

9.6 Four-Stage Methods

A general four-stage RK method has 10 free parameters ($\frac{1}{2}s(s+1)$) and eight conditions are needed to obtain methods of maximum possible order—which is $p = 4$ (Lambert [44, pages 178–9]). The most popular of all RK methods (of any stage number) is the four-stage, fourth-order method shown in Table 9.8; it is commonly referred to as *the* RK method.

0	0			
$\frac{1}{2}$	$\frac{1}{2}$	0		
$\frac{1}{2}$	0	$\frac{1}{2}$	0	
1	0	0	1	0
	$\frac{1}{6}$	$\frac{1}{3}$	$\frac{1}{3}$	$\frac{1}{6}$

Table 9.8 The classic four-stage, fourth-order method

9.7 Attainable Order of RK Methods

It may appear from the development to this point that it is always possible to find RK methods with s stages that have order s. This is not so for $s > 4$. The number of stages necessary for a given order is known up to order 8, but there are no precise results for higher orders. To appreciate the level of complexity involved, it is known that between 12 and 17 stages will be required for order 9 and the coefficients must satisfy 486 nonlinear algebraic equations. Also, for

methods of order higher than 4, the order when applied to systems may be
lower than when applied to scalar ODEs. A structure to tackle such problems
was provided by Butcher in the early 1970s, and many of the results that have
flowed from this are summarized in the books of Butcher [6] and Hairer et
al. [28].

9.8 Postscript

RK methods were devised in the early years of the 20th century. This was an
era when all calculations were performed by hand and so the free parameters
remaining, after the relevant order conditions were satisfied, were chosen to give
as many zero entries as possible in the Butcher array while trying to ensure
that the nonzero entries were fractions involving small integers. In the era of
modern digital computing, attention has shifted to minimizing the coefficients
that appear in the leading terms of the LTE.

EXERCISES

9.1.* Use the improved Euler RK(2) method with $h = 0.2$ to compute x_1
and x_2 for the IVP of Example 9.1 (this may be carried out with a
hand calculator).

9.2.* Prove that the RK method (9.5)–(9.6) applied to the IVP $x'(t) = 1$,
$x(0) = 0$, will not converge unless

$$\sum_{i=1}^{s} b_i = 1.$$

9.3.** As described in Section 1.1, the scalar ODE $x'(t) = f(t, x(t))$, with
$x(0) = \eta$, may be written in the form of two autonomous ODEs,
$\boldsymbol{u}'(t) = \boldsymbol{f}(\boldsymbol{u}(t))$, where

$$\boldsymbol{u}(t) = \begin{bmatrix} u(t) \\ v(t) \end{bmatrix}, \qquad \boldsymbol{f}\left(\begin{bmatrix} u \\ v \end{bmatrix}\right) = \begin{bmatrix} 1 \\ f(u, v) \end{bmatrix},$$

and $u(0) = 0$, $v(0) = \eta$. Show that the condition (9.7) arises naturally when we ask for a RK method to produce the same numerical
approximation when applied to either version of the problem.

9.4.* Show that the one-stage method in Section 9.3 is a special case of the two-stage process (9.9) by choosing appropriate values of a, b_1, and b_2. Hence show that we must have $b_1 = 1$ so that the method in Section 9.3 has order one.

9.5.** Prove, from first principles, that the RK method

$$k_1 = f(t_n, x_n),$$
$$k_2 = f(t_n + h, x_n + hk_1),$$
$$x_{n+1} = x_n + \tfrac{1}{2}h(k_1 + k_2),$$

is a consistent method (of order at least $p = 1$) for solving the IVP $x'(t) = f(t, x(t))$, with $x(0) = \eta$.

9.6.** Prove, from first principles, that the RK method

$$k_1 = f(t_n, x_n),$$
$$k_2 = f(t_n + \tfrac{3}{2}h, x_n + \tfrac{3}{2}hk_1),$$
$$x_{n+1} = x_n + \tfrac{1}{3}h(2k_1 + k_2),$$

is consistent of at least second order for solving the initial value problem $x'(t) = f(t, x(t))$, $t > 0$ with $x(0) = \eta$.

9.7.* Consider the solution of the IVP $x'(t) = (1 - 2t)x(t)$ with $x(0) = 1$. Verify the values (to six decimal places) given below for x_1 with $h = 0.1$ and the methods shown. The digits underlined coincide with those in the exact solution $x(h) = \exp[\tfrac{1}{4} - (\tfrac{1}{2} - h)^2]$.

Improved Euler	Modified Euler	Heun	Kutta third order	Fourth order
1.094 000	1.094 500	1.094 179	1.094 187	1.094 174

9.8.** Apply the general second-order, two-stage RK method (see Table 9.5) to the ODE $x'(t) = t^2$. Compare x_{n+1} with the Taylor expansion of $x(t_{n+1})$ and comment on the order of the LTE.

9.9.** Apply the general second-order, two-stage RK method (see Table 9.5) to the ODE $x'(t) = \lambda x(t)$. Compare x_{n+1} with the Taylor expansion of $x(t_{n+1})$ and show that the difference is $\mathcal{O}(h^3)$ if $x_n = x(t_n)$. We conclude that the order of the method cannot exceed two while the calculations in Section 9.4 show that the order is at least 2; the order must therefore be exactly 2.

9.10.*** Show that (9.4) follows from (9.3) (since $x_{n+1} = x_n + \frac{1}{2}h(k_1+k_2)$).

By matching coefficients of powers of $\widehat{h}$ on both sides of

$$1 - \tfrac{1}{2}\widehat{h} + \tfrac{1}{12}\widehat{h}^2 = (1 + \tfrac{1}{2}\widehat{h} + \tfrac{1}{12}\widehat{h}^2)(a_0 + a_1\widehat{h} + a_2\widehat{h}^3 + \cdots),$$

or otherwise, show that the Maclaurin expansion of the right side of (9.4) is

$$x_{n+1} = (1 + \widehat{h} + \tfrac{1}{2}\widehat{h}^2 + \tfrac{1}{6}\widehat{h}^3 + \tfrac{1}{24}\widehat{h}^4 + \tfrac{1}{144}\widehat{h}^5)x_n + \mathcal{O}(\widehat{h}^6).$$

Deduce that the method cannot be of order higher than four.

9.11.** Show that the RK method with Butcher tableau

$$
\begin{array}{c|cc}
0 & 0 & 0 \\
1 & \frac{1}{2} & \frac{1}{2} \\
\hline
 & \frac{1}{2} & \frac{1}{2}
\end{array}
$$

is equivalent to the trapezoidal rule (4.5).

9.12.* Find a solution of the third-order conditions so that $c_2 = c_3$ and $b_2 = b_3$ (Nyström's third-order method).

9.13.* Show that, for any given third-order, three-stage RK method, it is possible to construct a second-order, two-stage method that uses the same values of k_1 and k_2.

9.14.*** If $x'(t) = f(t, x(t))$, use the chain rule to verify that

$$x'''(t) = f_{tt} + 2ff_{xt} + f^2 f_{xx} + f_x(f_t + ff_x),$$

where f and its partial derivatives are evaluated at (x, t).

9.15.*** Use the result derived in the previous exercise to show, from first principles, that Heun's method (see Table 9.7) is a third-order method.

10
Runge-Kutta Methods–II
Absolute Stability

10.1 Absolute Stability of RK Methods

The notion of absolute stability developed in Chapter 6 for LMMs is equally relevant to RK methods. Applying an RK method to the linear ODE $x'(t) = \lambda x(t)$ with $\Re(\lambda) < 0$, absolute stability requires that $x_n \to 0$ as $n \to \infty$.

We begin with the most general two-stage, second-order RK method discussed in Section 9.4.

Example 10.1

Investigate the absolute stability of the RK method with Butcher array

$$
\begin{array}{c|cc}
0 & 0 & \\
a & a & 0 \\
\hline
& 1-\theta & \theta
\end{array}
\qquad a = 1/(2\theta).
$$

Applying this method to the solution of the ODE $x'(t) = \lambda x(t)$ and writing

D.F. Griffiths, D.J. Higham, *Numerical Methods for Ordinary Differential Equations*,
Springer Undergraduate Mathematics Series, DOI 10.1007/978-0-85729-148-6_10,
© Springer-Verlag London Limited 2010

$\widehat{h} = \lambda h$, we find that[1]

$$hk_1 = hf(t_n, x_n) = h\lambda x_n = \widehat{h}x_n,$$
$$hk_2 = hf(t_n + ah, x_n + ahk_1),$$
$$= \widehat{h}(x_n + ahk_1) = \widehat{h}(1 + a\widehat{h})x_n.$$

Hence,

$$x_{n+1} = x_n + (1 - \theta)hk_1 + \theta hk_2$$
$$= \big(1 + \widehat{h}(1 + \theta a\widehat{h})\big)x_n$$

and, since $a\theta = \frac{1}{2}$, it follows that

$$x_{n+1} = R(\widehat{h})x_n, \tag{10.1}$$

where

$$R(\widehat{h}) = 1 + \widehat{h} + \tfrac{1}{2}\widehat{h}^2 \tag{10.2}$$

is known as the *stability function*[2] of the RK method. Equation (10.1) is a one-step difference equation having auxiliary equation (stability polynomial)

$$p(r) = r - \big(1 + \widehat{h} + \tfrac{1}{2}\widehat{h}^2\big).$$

Some remarks are in order:

 (i) The stability function is linear in r and has only one root $r = R(\widehat{h})$.

 (ii) The stability function is the same for *all* explicit two-stage, second-order RK methods—$R(\widehat{h})$ does not depend on the parameter a.

(iii) The root $r = R(\widehat{h})$ satisfies $R(\widehat{h}) = e^{\widehat{h}} + \mathcal{O}(\widehat{h}^{p+1}))$ with $p = 2$ for this family of second-order methods.

(iv) The solution of (10.1) will tend to zero as $n \to \infty$, regardless of the value of x_0, if, and only if, $|R(\widehat{h})| < 1$; this is the condition for absolute stability.

The interval of absolute stability is the set of real values $\widehat{h}$ for which $|R(\widehat{h})| < 1$. This leads to

$$-1 < 1 + \widehat{h} + \tfrac{1}{2}\widehat{h}^2 < 1.$$

The left inequality is satisfied for all real $\widehat{h}$ and the right inequality is satisfied if, and only if, $-2 < \widehat{h} < 0$. This is the interval of stability. The boundary of the

---

[1]It is usually more convenient to calculate hk_i (for each i) since the result can be expressed in terms of $\widehat{h}$ only, and does not involve h or λ separately.

[2]Clearly, $R(\widehat{h})$ is a polynomial in this example and is, indeed, a polynomial for all explicit methods. However, as seen in Example 9.2, implicit methods lead to rational expressions in $\widehat{h}$.

region of absolute stability, despite having a rather a simple shape (Figure 10.1, top right), is defined by a complicated expression (Exercise 10.9). What is required is a means of computing the largest value of h for a given (complex) value of λ. Writing $\lambda = a + ib$, then $\widehat{h} = h(a + ib)$ and it can be shown that the equation $|1 + \widehat{h} + \frac{1}{2}\widehat{h}^2|^2 = 1$ leads to the following cubic polynomial in h:

$$(a^2 + b^2)^2 h^3 + 4a(a^2 + b^2)h^2 + 8a^2h + 8a = 0. \tag{10.3}$$

The real root of this equation gives the largest value of h for which the method is absolutely stable. □

10.1.1 s-Stage Methods of Order s

The results of Example 10.1 generalize to s-stage methods of order s (which exist only for $s \leq 4$—see Section 9.7) as follows:

(a) When applied to $x'(t) = \lambda x(t)$, the RK method leads to $x_{n+1} = R(\widehat{h})x_n$, where the stability function $R(\widehat{h})$ is a polynomial in $\widehat{h}$ of degree s (see Exercises 10.7 and 10.8).

(b) Taylor expansion of the exact solution $x(t_{n+1})$ about $t = t_n$ and using $x'(t) = \lambda x(t)$, $x''(t) = \lambda^2 x(t)$, etc., leads to

$$x(t_{n+1}) = \left(1 + \widehat{h} + \frac{1}{2!}\widehat{h}^2 + \cdots + \frac{1}{s!}\widehat{h}^s\right)x(t_n) + \mathcal{O}(\widehat{h}^{s+1}). \tag{10.4}$$

(c) Under the *localizing assumption* $x_n = x(t_n)$ and supposing the method to have order s, then $x_{n+1} = x(t_{n+1}) + \mathcal{O}(h^{s+1})$. It follows from (a) and (b) that

$$R(\widehat{h}) = 1 + \widehat{h} + \frac{1}{2!}\widehat{h}^2 + \cdots + \frac{1}{s!}\widehat{h}^s. \tag{10.5}$$

Thus, all s-stage, s-order RK methods have the same stability function, $R(\widehat{h})$. Clearly,

$$R(\widehat{h}) = e^{\widehat{h}} + \mathcal{O}(\widehat{h}^{s+1}).$$

Also $|R(\widehat{h})| \to \infty$ as $\widehat{h} \to -\infty$ so that no explicit RK method can be either A-stable or A_0-stable. The intervals/regions of absolute stability are defined as the set of real/complex values of $\widehat{h}$ for which $|R(\widehat{h})| < 1$. (Since $R(\widehat{h}) = 1$ at $\widehat{h} = 0$, RK methods are always zero-stable.)

The intervals of absolute stability (IAS) of the four s-stage, sth-order RK methods ($1 \leq s \leq 4$) are given in Table 10.1 and their regions of absolute stability are shown in Figure 10.1. Note that the area of the region, as well as the length of the interval of absolute stability, increases with s.

s	$R(\widehat{h})$	IAS
1	$1 + \widehat{h}$	$(-2, 0)$
2	$1 + \widehat{h} + \frac{1}{2}\widehat{h}^2$	$(-2, 0)$
3	$1 + \widehat{h} + \frac{1}{2}\widehat{h}^2 + \frac{1}{6}\widehat{h}^3$	$(-2.513, 0)$
4	$1 + \widehat{h} + \frac{1}{2}\widehat{h}^2 + \frac{1}{6}\widehat{h}^3 + \frac{1}{24}\widehat{h}^4$	$(-2.785, 0)$

Table 10.1 Intervals of absolute stability (IAS) for s-stage, s-order explicit RK methods

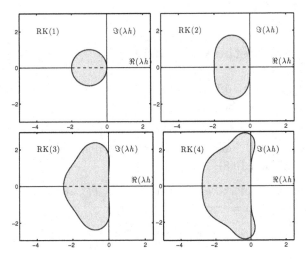

Fig. 10.1 The region of of absolute stability for the s-stage, sth-order RK methods. (RK(1) is Euler's method: cf. Figure 6.6)

10.2 RK Methods for Systems

When applying explicit RK methods to IVPs for systems of the form

$$\left.\begin{array}{l} \boldsymbol{x}'(t) = \boldsymbol{f}(t, \boldsymbol{x}(t)), \quad t > t_0 \\ \boldsymbol{x}(t_0) = \boldsymbol{\eta} \end{array}\right\}, \tag{10.6}$$

where $\boldsymbol{x} \in \mathbb{R}^d$ is a d-dimensional vector, each of the stage slopes $\boldsymbol{k}_1, \boldsymbol{k}_2, \ldots, \boldsymbol{k}_s$ becomes a vector. When performing hand calculations it may be preferable to use a different symbol for each component of $\boldsymbol{k}_i$:

$$\boldsymbol{k}_1 = \begin{bmatrix} k_1 \\ \ell_1 \\ \vdots \end{bmatrix}, \quad \boldsymbol{k}_2 = \begin{bmatrix} k_2 \\ \ell_2 \\ \vdots \end{bmatrix}, \quad \text{etc.}$$

This is illustrated in the next example, which revisits Example 7.1.

Example 10.2

Use the improved Euler RK(2) method with $h = 0.1$ to determine the solution at $t = 0.1$ of the IVP

$$u'(t) = -tu(t)v(t),$$
$$v'(t) = -u^2(t),$$

with $u = 1$, $v = 2$ at $t = 0$.

The Butcher array of the improved Euler method is (Table 9.5 with $\theta = \frac{1}{2}$)

$$
\begin{array}{c|cc}
0 & 0 & \\
\frac{1}{2} & \frac{1}{2} & 0 \\
\hline
 & \frac{1}{2} & \frac{1}{2}
\end{array}
$$

$n = 0$: $t_0 = 0$,

$k_1 = -t_0 u_0 v_0 = 0$,

$\ell_1 = -u_0^2 = -1$,

$k_2 = -(t_0 + \frac{1}{2}h)(u_0 + \frac{1}{2}hk_1)(v_0 + \frac{1}{2}h\ell_1) = -0.0975$,

$\ell_2 = -(u_0 + \frac{1}{2}hk_1)^2 = -1$,

$u_1 = u_0 + h(k_1 + k_2) = 0.9902$,

$v_1 = v_0 + h(\ell_1 + \ell_2) = 1.900$,

$t_1 = 0.1$.

Note that u gets updated using k and the v gets updated using ℓ. A possible layout of calculations is shown in Table 10.2, where they are continued for one further step.

n	t_n	k_1	k_2	$x_n = \begin{bmatrix} u_n \\ v_n \end{bmatrix}$	
0	0.0			1	
				2	Initial data
1	0.1	0.0000	−0.0975	0.9902	
		−1.0000	−1.0000	1.9000	
2	0.2	−0.1881	−0.2723	0.9630	
		−0.9806	−0.9621	1.8038	

Table 10.2 Numerical solutions for the first two steps of Example 10.2

10.3 Absolute Stability for Systems

The absolute stability of LMMs applied to systems was discussed in Section 7.1—see Definition 7.4; a similar investigation is now carried out for RK methods by applying them to the model problem

$$u'(t) = Au(t), \tag{10.7}$$

where $u(t)$ is an m-dimensional vector $(u(t) \in \mathbb{R}^m)$ and A is a constant $m \times m$ matrix $(A \in \mathbb{R}^{m \times m})$.

Example 10.3

Investigate the absolute stability properties of the family of second-order, two-stage RK(2) methods.

The scalar version of this example is treated in Example 10.1. The corresponding calculations are

$$
\begin{aligned}
hk_1 &= hf(t_n, u_n) = hAu_n, \\
hk_2 &= hf(t_n + ah, u_n + ahk_1) \\
&= hA(u_n + ahk_1) = hA(1 + ahA)u_n.
\end{aligned}
$$

Hence,

$$
\begin{aligned}
u_{n+1} &= u_n + (1 - \theta)hk_1 + \theta hk_2 \\
&= \left(1 + hA + \tfrac{1}{2}hA^2\right)u_n
\end{aligned}
$$

(since $a\theta = \tfrac{1}{2}$), so that

$$u_{n+1} = R(hA)u_n, \tag{10.8}$$

where $R(\widehat{h})$ is the stability function defined by equation (10.2). Thus, the matrix-vector version is analogous to the scalar version provided that $\widehat{h}(= h\lambda)$ is interpreted here as hA.

We assume, as in Section 7.1, that A is diagonalized by the matrix V, so that $V^{-1}AV = \Lambda$, where Λ is the diagonal matrix of eigenvalues. Then

$$V^{-1}R(hA)V = R(h\Lambda)$$

(see Exercise 10.12). Multiplying both sides of (10.8) by V^{-1} and applying the change of variables $u_{n+j} = Vx_{n+j}$ leads to

$$x_{n+1} = R(h\Lambda)x_n,$$

a system in which all the components are uncoupled. If x_n denotes a typical component of $\boldsymbol{x}_n$ and λ the corresponding eigenvalue, then

$$x_{n+1} = R(\widehat{h})x_n, \qquad \widehat{h} = h\lambda,$$

the same equation that was obtained in the scalar case.

The results of this example generalize quite naturally to other explicit RK methods; to ensure absolute stability, $\widehat{h} = h\lambda$ should lie in the region of absolute stability of the method for every eigenvalue λ of A. $\qquad\qquad\square$

EXERCISES

10.1.* Verify that the two three-stage RK methods given in Table 9.7 have the same stability function $R(\widehat{h})$.

10.2.** Show, by considering the roots of the equations $R(\widehat{h}) = 1$ and $R(\widehat{h}) = -1$, that all three-stage, third-order RK methods have the same interval of absolute stability $(h^*, 0)$, where h^* lies between -2 and -3.

10.3.*** Apply the RK(4) method defined in Table 9.8 to the ODE $x'(t) = \lambda x(t)$ and verify that it leads to $x_{n+1} = R(\widehat{h})x_n$, where $R(\widehat{h})$ is defined by (10.5) with $s = 4$.

Show that $R(\widehat{h})$ may be written as

$$R(\widehat{h}) = \tfrac{1}{4} + \tfrac{1}{3}(\widehat{h} + \tfrac{3}{2})^2 + \tfrac{1}{24}\widehat{h}^2(\widehat{h} + 2)^2$$

and deduce that the equation $R(\widehat{h}) = -1$ has no real roots.

Prove that the method has interval of absolute stability $(h^*, 0)$, where $R(h^*) = 1$, and h^* lies between -2 and -3.

10.4.** Show that s-stage, sth-order RK methods have interval of stability $(h^*, 0)$ for $s = 1, 2, 3, 4$, where h^* is defined by $R(h^*) = (-1)^s$.

10.5.** Use the Newton–Raphson method to solve the equations $R(h^*) = (-1)^s$ for $s = 3$ and $s = 4$ obtained in the previous exercise. Use a starting guess $h^{[0]} = -2.5$ and verify that the roots agree with the values given in Table 10.1 to three decimal places.

10.6.** Determine the interval of absolute stability for the RK method

$$k_1 = f(t_n, x_n),$$
$$k_2 = f(t_n + h/a, x_n + hk_1/a),$$
$$x_{n+1} = x_n + h((1-a)k_1 + ak_2)$$

and show that it is independent of the parameter a.

Show how the stability function for this method can be used to establish an upper limit to its order of convergence.

10.7.*** Suppose that an RK method, defined by the Butcher array

$$\begin{array}{c|c} \mathbf{c} & \mathcal{A} \\ \hline & \mathbf{b}^{\mathrm{T}} \end{array}$$

with $\mathbf{c} = (c_1, \ldots, c_s)^{\mathrm{T}}$, $\mathbf{b} = (b_1, \ldots, b_s)^{\mathrm{T}}$ and $\mathcal{A} = (a_{i,j})$, an $s \times s$ matrix, is applied to the scalar test equation $x'(t) = \lambda x(t)$. If $\mathbf{k} = [k_1, \ldots, k_s]^{\mathrm{T}}$ and $\mathbf{e} = [1, 1, \ldots, 1]^{\mathrm{T}}$, show that

$$\mathbf{k} = \lambda(I - \widehat{h}\mathcal{A})^{-1}\mathbf{e}x_n.$$

Hence show that $x_{n+1} = R(\widehat{h})x_n$ and the stability function $R(\widehat{h})$ is given by

$$R(\widehat{h}) = 1 + \widehat{h}\mathbf{b}^{\mathrm{T}}(I - \widehat{h}A)^{-1}\mathbf{e}.$$

10.8.*** Using the Cayley–Hamilton Theorem, or otherwise, prove that $\mathcal{A}^s = 0$ (it is nilpotent) when $\mathcal{A}$ is a strictly lower triangular $s \times s$ matrix . Hence prove that

$$(I - \widehat{h}\mathcal{A})^{-1} = I + \widehat{h}\mathcal{A} + \widehat{h}^2\mathcal{A}^2 + \cdots + \widehat{h}^{s-1}\mathcal{A}^{s-1}.$$

Deduce that, when $\mathcal{A}$ is the Butcher matrix for an explicit RK method, the stability function $R(\widehat{h})$ of Exercise 10.7 is a polynomial in $\widehat{h}$ of degree s.

10.9.*** The boundary of the region of absolute stability of all second-order two-stage RK methods is given by $|1 + \widehat{h} + \frac{1}{2}\widehat{h}^2| = 1$. If $\widehat{h} = p + iq$, show that this leads to

$$(p + 1)^2 + \left(\sqrt{q^2 + 1} - 1\right)^2 = 1.$$

Hence, show that the boundary can be parameterized in real form by

$$p = \cos(\phi) - 1, \quad q = \pm\sqrt{(2 + \sin(\phi))\sin(\phi)}$$

for $0 \le \phi \le \pi$.

10.10.* If $\lambda = -4 \pm i$ show, using (10.3), that all two-stage, second-order RK methods are absolutely stable for $0 < h < 0.498$.

10.11.** Show that any two-stage, second-order RK method applied to the system $\boldsymbol{u}'(t) = A\boldsymbol{u}(t)$, where

$$A = \begin{bmatrix} -5 & -2 \\ 2 & -5 \end{bmatrix},$$

will be absolutely stable if $h < 0.392$. Hint: use (10.3).

What is the corresponding result for the matrix $A = \begin{bmatrix} -50 & -20 \\ 20 & -50 \end{bmatrix}$?

10.12.** Suppose that the $s \times s$ matrix A is diagonalized by the $s \times s$ matrix V :

$$V^{-1}AV = \Lambda,$$

where Λ is an $s \times s$ diagonal matrix. Prove that V also diagonalizes all positive powers of A, i.e. $V^{-1}A^k V = \Lambda^k$ for any positive integer k.

Deduce that $V^{-1}R(hA)V = R(h\Lambda)$, where $R(\widehat{h})$ is a polynomial of degree s in $\widehat{h}$.

10.13.*** Prove that the method of Example 9.2 is A-stable, i.e. absolutely stable for all $\Re(\widehat{h}) < 0$.

[Hint: Write $R(\widehat{h}) = 1 + \widehat{h}/D$ and show that $|R(\widehat{h})|^2 - 1 < 0$.]

10.14.*** Show that the RK method given by the Butcher matrix

$$\begin{array}{c|ccc}
0 & 0 & & \\
c & c & 0 & \\
\hline
 & c & 1-c &
\end{array}$$

is consistent of order 1.

Show that one step of this method is equivalent to taking two steps with Euler's method—the first step with a step size ch and the second with a step size $(1-c)h$. When $c = \frac{1}{2}(1-\gamma)$, relate this to the composite Euler method described in Exercise 6.21 and, hence, deduce the interval of absolute stability of the given RK method.

10.15.*** A semi-implicit RK method is given by the Butcher matrix

$$\begin{array}{c|ccc}
0 & 0 & 0 & 0 \\
\frac{1}{2} & \frac{5}{24} & \frac{1}{3} & -\frac{1}{24} \\
1 & 0 & 1 & 0 \\
\hline
 & \frac{1}{6} & \frac{2}{3} & \frac{1}{6}
\end{array}$$

Determine the ratio x_{n+1}/x_n when the method is applied to $x'(t) = \lambda x(t)$. Deduce that the method cannot be A-stable.

11

Adaptive Step Size Selection

All the methods discussed thus far have been parameterized by the step size h. The number of steps required to integrate over a given interval $[0, t_f]$ is proportional to $1/h$ and the accuracy of the results is proportional to h^p, for a method of order p. Thus, halving h is expected to double[1] the amount of computational effort while reducing the error by a factor of 2^p (more than an extra digit of accuracy if $p > 3$).

We now explore the possibility of taking a different step length, h_n, at step number n, say, in order to improve efficiency—to obtain the same accuracy with fewer steps or better accuracy with the same number of steps. We want to *adapt* the step size to local conditions—to take short steps when the solution varies rapidly and longer steps when there is relatively little activity. The process of calculating suitable step sizes should be automatic (by formula rather than by human intervention) and inexpensive (accounting for a small percentage of the overall computing cost, otherwise one might as well repeat the calculations with a smaller, constant step size h).

We shall describe methods for computing numerical solutions at times $t = t_0, t_1, t_2, \ldots$ that are not equally spaced, so we define the sequence of step sizes

$$h_n = t_{n+1} - t_n, \quad n = 0, 1, 2, \ldots. \tag{11.1}$$

What should be the strategy for calculating these step sizes? It appears intuitively attractive that they be chosen so as to ensure that our solutions

---

[1] These ball-park estimates are based on h being sufficiently small so that $\mathcal{O}(h^p)$ quantities are dominated by their leading terms.

D.F. Griffiths, D.J. Higham, *Numerical Methods for Ordinary Differential Equations*, Springer Undergraduate Mathematics Series, DOI 10.1007/978-0-85729-148-6_11, © Springer-Verlag London Limited 2010

have a certain accuracy—three correct decimal places or four correct significant figures, for example. This is not always possible, as the following example
illustrates.

Example 11.1

Using knowledge of the exact solution of the IVP

$$x'(t) = 2t, \qquad x(0) = 0,$$

show that it is not possible to choose the sequence of step sizes h_n in Euler's
method so that the solution is correct to precisely 0.01 at each step.

The method is given by

$$x_{n+1} = x_n + 2h_n t_n,$$
$$t_{n+1} = t_n + h_n,$$

with $x_0 = t_0 = 0$. Thus, $x_1 = 0$, $t_1 = h_0$, while the exact solution at this time
is $x(t_1) = h_0^2$. The GE $x(t_1) - x_1$ is equal to 0.01 when $h_0^2 = 0.01$; that is,
$h_0 = 0.1$. For the second step,

$$x_2 = x_1 + 2h_1 t_1,$$

where $t_1 = 0.1$, so $x_2 = 0.2h_1$ while $x(t_2) = (0.1 + h_1)^2$. The GE is

$$x(t_1) - x_1 = 0.01 + h_1^2$$

and can equal 0.01 only if $h_1 = 0$. It is not possible to obtain a GE of 0.01 after
two steps unless the GE after one step is less than this amount.   □

Thus, in order to obtain a given accuracy at time $t = t_f$, say, it may be necessary in some problems to have much greater accuracy at earlier times; in others,
lower accuracy may be sufficient. While it is possible to devise algorithms that
can cope with both scenarios (see, for instance, Eriksson et al. [19]), most current software for solving ODEs is not based on controlling the GE. Rather, the
most common strategy is to require the user to specify a number tol, called the
tolerance, and the software is designed to adapt the step size so that the LTE
at each step does not exceed this value.[2] Our aim here is to present a brief
introduction to this type of adaptive step size selection. We shall address only
one step methods, in which case the calculations are based on the relationship

$$T_{n+1} = H(t_n)h_n^{p+1} + \mathcal{O}(h_n^{p+2}) \tag{11.2}$$

---

[2]The situation is a little more intricate in practice, since software is typically
designed to use a relative error tolerance in addition to an absolute error tolerance—
see, for instance, the book by Shampine et al. [63, Section 1.4]. We shall limit ourselves
to just one tolerance.

(for some function $H(t)$) between the step size h_n and the leading term in the LTE of a pth-order method[3] for the step from $t = t_n$ to $t = t_{n+1}$.

The calculations are organized along the following lines. Suppose that the pairs (t_0, x_0), (t_1, x_1), ..., (t_n, x_n) have already been computed for some value of $n \geq 0$ along with a provisional value h_{new} for h_n, the length of the next time step. There are four main stages in the calculation of the next step.

(a) Set $h_n = h_{\text{new}}$ and calculate provisional values of x_{n+1} and $t_{n+1} = t_n + h_n$.

(b) Estimate a numerical value $\widehat{T}_{n+1}$ for the LTE T_{n+1} at $t = t_{n+1}$ based on data currently available. (How this is done is method-specific and will be described presently.)

(c) If $\widehat{T}_{n+1} < \text{tol}$ we could have used a larger step size, while if $\widehat{T}_{n+1} > \text{tol}$ the step we have taken is unacceptable and it will have to be recalculated with a smaller step size.

In both situations we ask: "Based on (11.2), what step size, h_{new}, would have given an LTE exactly equal to tol?" From step (b) we know that $\widehat{T}_{n+1} \approx H(t_n) h_n^{p+1}$, while to achieve $|T_{n+1}| = \text{tol}$ in (11.2) we require

$$\text{tol} \approx |H(t_n)| h_{\text{new}}^{p+1}; \tag{11.3}$$

so, after eliminating $H(t_n)$, we take h_{new} to be

$$h_{\text{new}} = h_n \left| \frac{\text{tol}}{\widehat{T}_{n+1}} \right|^{1/(p+1)}. \tag{11.4}$$

This is used in two ways in the next stage.

(d) If the estimated value of LTE obtained in stage (b) is too large[4]—$|T_{n+1}| > \text{tol}$—the current step is rejected. In this case we return to the previous step—stage (a) on this list—with the value of h_{new} given by (11.4) and recalculate x_{n+1} and t_{n+1}.

Otherwise we proceed to the next time step with the values of (t_{n+1}, x_{n+1}) and h_{new}.

This process is common to all the methods we will describe and its implementation requires a means of calculating an estimate $\widehat{T}_{n+1}$ of the LTE (T_{n+1}) and a suitable initial step length (h_0).

---

[3]For instance, $T_{n+1} = \mathscr{L}_h x(t_n) = C_{p+2} h_n^{p+1} x^{(p+1)}(t_n) + \mathcal{O}(h^{p+2})$ for a pth-order LMM (see Equation (4.18)) so $H(t) = C_{p+1} x^{(p+1)}(t)$. The corresponding expression for two-stage RK methods is given at the end of Section 9.4, but, as we shall see, the function $H(t)$ for RK methods is estimated indirectly.

[4]In practice a little leeway is given and a step is rejected if, for example, $|\widehat{T}_{n+1}| > 1.1 \text{tol}$.

Equation (11.4) implies that $h_{\text{new}} \propto \text{tol}^{1/(p+1)}$ and, since the GE is of order p, it follows that $\text{GE} \propto \text{tol}^{p/(p+1)}$. For this reason we conduct numerical experiments in pairs with tolerances that differ by a ratio of 10^{p+1}. The size of computed time steps should then differ by a factor of about 10 and experiments with the smaller tolerance should require about 10 times as many steps to integrate over the same interval. Also, the GE in the two experiments should differ by a factor of 10^p, so the smaller tolerance should have p more correct decimal digits than the larger tolerance. This is often made clearer graphically by plotting the *scaled* GE, defined by

$$\text{scaled GE} = \frac{\text{GE}}{\text{tol}^{p/(p+1)}},$$

as a function of time. The curves produced by the two experiments should lie roughly on top of each other.

In the remainder of this chapter we will present a number of examples to illustrate how methods of different types may be constructed and, via numerical examples, give some indication of how they perform.

11.1 Taylor Series Methods

These are the most straightforward methods for which to design step size selection, since the leading term in the LTE may be expressed in terms of $x(t)$ and t by forming successive derivatives of the ODE.

The TS(p) method is given by (see equation (3.5))

$$x_{n+1} = x_n + h_n x_n' + \tfrac{1}{2!} h_n^2 x_n'' + \cdots + \tfrac{1}{p!} h_n^p x_n^{(p)}. \tag{11.5}$$

The LTE for the step from $t = t_n$ to $t = t_{n+1}$ is given by

$$T_{n+1} = \tfrac{1}{(p+1)!} h_n^{p+1} x^{(p+1)}(t_n) + \mathcal{O}(h_n^{p+2}),$$

in which the leading term can be approximated by replacing $x^{(p+1)}(t_n)$ by $x_n^{(p+1)}$. Thus,

$$\widehat{T}_{n+1} = \tfrac{1}{(p+1)!} h_n^{p+1} x_n^{(p+1)}, \tag{11.6}$$

whose right-hand side is a computable quantity since it can be expressed in terms of the previously computed quantities h_n, t_n, and x_n by differentiating the ODE p times. With (11.4) and (11.6) the step size update is given by

$$h_{\text{new}} = \left| \frac{(p+1)!\, \text{tol}}{x_n^{(p+1)}} \right|^{1/(p+1)}; \tag{11.7}$$

so, if the current step is accepted, we progress with $h_{n+1} = h_{\text{new}}$. Otherwise the step is recomputed with $h_n = h_{\text{new}}$.

It remains to choose a suitable value for the first step length h_0. This may be done by setting $|T_0| = \text{tol}$, from which we find

$$h_0 = \left| \frac{(p+1)!\,\text{tol}}{x^{(p+1)}(t_0)} \right|^{1/(p+1)}. \tag{11.8}$$

Example 11.2

Apply the TS(1) and TS(2) methods to solve the IVP (see Example 2.1)

$$\left. \begin{array}{l} x'(t) = (1 - 2t)x(t), \quad t > 0 \\ x(0) = 1 \end{array} \right\} \tag{11.9}$$

using adaptive step size selection.

The TS(1) method at the nth step is Euler's method with a step size h_n:

$$x_{n+1} = x_n + h_n x_n', \qquad t_{n+1} = t_n + h_n, \tag{11.10}$$

where $x_n' = (1 - 2t_n)x_n$. By differentiating the ODE (see Example 2.1) we find

$$x''(t_n) = \left[(1 - 2t_n)^2 - 2 \right] x(t_n)$$

and this may be approximated by replacing the exact solution $x(t_n)$ by the numerical solution x_n at $t = t_n$ to give

$$x_n'' = \left[(1 - 2t_n)^2 - 2 \right] x_n. \tag{11.11}$$

The step size update is then given by (11.4) with $p = 1$; that is,[5]

$$h_{\text{new}} = h_n \left| \frac{2\,\text{tol}}{x_n''} \right|^{1/2}. \tag{11.12}$$

Finally, since $t_0 = 0$ and $x_0 = 1$ we find from (11.11) that $x_0'' = -1$ and so equation (11.8) gives $h_0 = \sqrt{2\,\text{tol}}$.

---

[5]The denominator in (11.12) vanishes when $x_{n+1} = 0$, but this is easily accommodated since Equation (11.10) then implies that $f_{n+1} = 0$ so that $x_{n+2} = x_{n+1} = 0$ at all subsequent steps.

A more significant problem would occur at $t_n = \frac{1}{2}(1 + \sqrt{2})$, since it would lead to $x_n'' = 0$. When t_n is close to this value the leading term in in T_{n+1} may well be much smaller than the remainder and our estimate of the LTE is likely to be too small. Should this lead to h_{new} being too large the check carried out at the end of each step would reject the step and it would be recomputed with a smaller step size. If it should happen that $x_n'' = 0$, division by zero is avoided in our experiments by choosing $h_{n+1} = h_n$.

	TS(1)			TS(2)	
tol	No. steps	Max. error	tol	No. steps	Max. error
10^{-2}	21	8.3×10^{-2}	10^{-2}	11	2.3×10^{-2}
10^{-4}	207	8.4×10^{-3}	10^{-5}	101	1.8×10^{-4}

Table 11.1 Numerical results for TS(1) and TS(2) methods for solving the IVP (11.9) in Example 11.2

The results of numerical experiments with tol $= 10^{-2}$ and 10^{-4} are reported in Table 11.1 and Figure 11.1. The discussion on page 148, just before the start of this section, suggests that $h_{\text{new}} \propto \text{tol}^{1/2}$; so, reducing tol by a factor of 100 should result in around 10 times as many time steps being required. This is confirmed by the second column of the table, where the number of steps increases from 21 to 207. The results in the third column of the table show that the GE is reduced (almost exactly) by a factor of 10 when tol is reduced by a factor of 100, confirming the expected relationship GE $\propto \text{tol}^{1/2}$.

In the time step history shown in Figure 11.1 (right) there is a local increase in h_n for $t \approx 1.2$ corresponding to the zero in the denominator of (11.12) (see footnote 5). It does not appear from the left-hand figure to have any significant effect on the GE. The asterisks indicate rejected time steps (see stage (d) on page 147). An explanation for the rejection of the final step in each of the simulations is offered in Exercise 11.4.

For the second-order method TS(2) (see Example 3.1)

$$x_{n+1} = x_n + h_n x_n' + \tfrac{1}{2} h_n^2 x_n'', \qquad t_{n+1} = t_n + h_n, \qquad (11.13)$$

and the step size is updated using (11.7) with $p = 2$. The third derivative

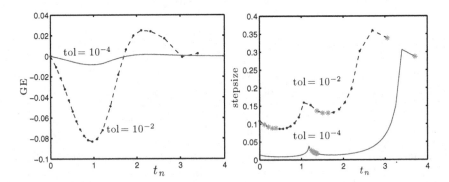

Fig. 11.1 Numerical results for Example 11.2 using TS(1). Shown are the variation of GE (left) and time step h_n (right) versus time t_n with tolerances tol $= 10^{-2}, 10^{-4}$

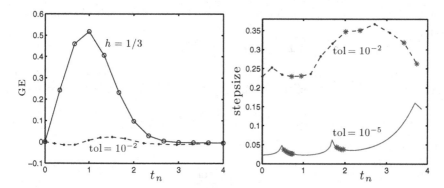

Fig. 11.2 Numerical results for Example 11.2 using TS(2). Left: the variation of GE for the adaptive method with tol $= 10^{-2}$ (solid dots) and also the fixed time step method with $h = 1/3$ (circles). Right: time step h_n versus time t_n with tolerances tol $= 10^{-2}, 10^{-5}$. Asterisks ($*$) indicate rejected steps

required can be found by differentiating the expression given earlier for $x''(t)$.

The results of numerical experiments with tol $= 10^{-2}$ and 10^{-5} shown in Table 11.1 are in accordance with the estimates $h_n \propto \text{tol}^{1/3}$ and GE $\propto \text{tol}^{2/3}$. It is also evident that TS(2) achieves much greater accuracy than TS(1) with substantially fewer steps.

The GE as a function of time is shown in Figure 11.2 (left) for tol $= 10^{-2}$. Also shown is the GE of the fixed-step version of TS(2) described in Example 3.2 that uses the same number of steps (12); thus, $h = 1/3$. The peak GE of the fixed-step method is more than 20 times larger than the variable-step equivalent, illustrating the significant gains in efficiency that can be achieved with the use of adaptive time steps.   □

Our final example of the TS method illustrates how the same ideas can be applied to the solution of systems of ODEs. It also serves as a warning that time steps have to conform to the absolute stability requirements of the method.

Example 11.3

Use the TS(1) method with automatic time step selection to solve the IVP (1.14) introduced in Example 1.9 to describe a cooling cup of coffee.

The system (1.14) can be written in the matrix-vector form $x'(t) = Ax(t) + g$, where

$$x(t) = \begin{bmatrix} u(t) \\ v(t) \end{bmatrix}, \qquad A = \begin{bmatrix} -8 & 8 \\ 0 & -1/8 \end{bmatrix}, \qquad g = \begin{bmatrix} 0 \\ 5/8 \end{bmatrix},$$

and, since g is constant, $x''(t) = Ax'(t) = A^2x(t) + Ag$. The initial condition is $\boldsymbol{x}(0) = [100, 20]^{\mathrm{T}}$. The equations in (11.10) have their obvious vector counterparts:

$$\boldsymbol{x}_{n+1} = \boldsymbol{x}_n + h_n \boldsymbol{x}'_n, \qquad t_{n+1} = t_n + h_n; \qquad (11.14)$$

to interpret the vector form of (11.12), we need a means of measuring the magnitude of the vector $\boldsymbol{x}''_n$. We shall use the Euclidean length, defined by

$$\|\boldsymbol{x}\| = (\boldsymbol{x}^{\mathrm{T}} \boldsymbol{x})^{1/2}$$

for a real vector $\boldsymbol{x}$. Hence, (11.12) becomes

$$h_{\mathrm{new}} = h_n \left(\frac{2\,\mathsf{tol}}{\|\boldsymbol{x}''_n\|} \right)^{1/2}. \qquad (11.15)$$

The initial time step is given by the vector analogue of (11.8); that is,

$$h_0 = \left(\frac{2\,\mathsf{tol}}{\|\boldsymbol{x}''_0\|} \right)^{1/2},$$

which gives $h_0 \approx 0.0198 \times \mathsf{tol}^{1/2}$. This, together with (11.14) and (11.15), serves to define the method.

The numerical results obtained with this method for $\mathsf{tol} = 10^{-2}$ and 10^{-4} are shown in Figure 11.3. The GE shown on the left appears to behave as expected—reducing tol by a factor of 100 reduces the GE by a factor of 10. To check this more carefully we have plotted (middle) the GE scaled by $\mathsf{tol}^{1/2}$. The two sets of results lie on top of each other for early times (up to about $t = 2$), but there is evident divergence in the later stages, accompanied by perceptible oscillations for $\mathsf{tol} = 10^{-2}$.

This departure from expected behaviour is due to the fairly violent oscillations in the time step, as seen in the rightmost plot. These, in turn, result from the need to restrict the step sizes in order to achieve absolute stability (see Chapter 6). To explain this, we first note that the eigenvalues of the matrix

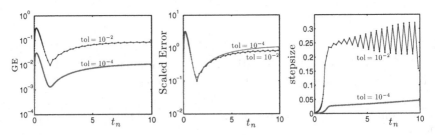

Fig. 11.3 Numerical results for Example 11.3 using TS(1). Shown are the variation of GE (left), the scaled GE (that is, GE/$\mathsf{tol}^{1/2}$) (middle), and time step h_n (right) versus time t_n. The tolerances used are $\mathsf{tol} = 10^{-2}, 10^{-4}$

A—which govern the dynamics of the system—are $\lambda = -8$ and $-1/8$ and the interval of absolute stability of Euler's method (Example 6.7) is $(-2, 0)$. It is therefore necessary for $\widehat{h} \equiv \lambda h \in (-2, 0)$ for each eigenvalue, that is, $h \leq \frac{1}{4}$.

Whenever formula (11.15) returns a value $h_{\mathrm{new}} > \frac{1}{4}$, instability causes an increase in the local error. The step size is then reduced below this critical level at the next step. The LTE is then smaller than tol, inducing an increase in step size, and the oscillations therefore escalate with time.

Because the magnitude of the solution in this example is initially 100, compared with 1 in previous examples, these tolerances are, relatively speaking, 100 times smaller than previously. Equivalent tolerances in this example would need to be 100 times larger than those we have used, but these would have led to even more severe oscillations in the time step. With tol $= 10^{-2}$ the level of GE is roughly 0.1, which corresponds to a relative error of only 0.1%—so, meeting the requirements of absolute stability leads to perhaps smaller GEs than are strictly necessary for most applications.

The moral of this example is that methods with step size control cannot overcome the requirements of absolute stability, and these requirements may force smaller step sizes—and hence more computational effort and higher accuracy—than the choice of tol would suggest.   □

11.2 One-Step Linear Multistep Methods

The general procedure for selecting time steps for LMMs is essentially the same as that for TS methods described in the previous section. The main difference is that repeated differentiation of the differential equation cannot be used to estimate the LTE, as this would negate the benefits of using LMMs. Hence, a new technique has to be devised for estimating higher derivatives $x''(t_n)$, $x'''(t_n)$, etc. of the exact solution.

The study of k-step LMMs with $k > 1$ becomes quite involved, so we will restrict ourselves to the one-step case.

Example 11.4

Devise a strategy for selecting the step size in Euler's method (TS(1)) that does not require differentiation of the ODE. Illustrate by applying the method to the IVP (11.9) used in Example 11.2.

The underlying method is the same as that in Example 11.2, namely

$$x_{n+1} = x_n + h_n x_n', \qquad t_{n+1} = t_n + h_n,$$

for which the LTE at the end of the current step is

$$T_{n+1} = \tfrac{1}{2!} h_n^2 x''(t_n) + \mathcal{O}(h_n^2).$$

The term $x''(t_{n+1})$ is approximated using the Taylor expansion of $x'(t_{n+1})$ about the point t_n:

$$x'(t_{n+1}) = x'(t_n) - h_n x''(t_n) + \mathcal{O}(h_n^2),$$

which can be rearranged to give

$$x''(t_n) = \frac{x'(t_{n+1}) - x'(t_n)}{h_n} + \mathcal{O}(h_n).$$

This can be estimated in terms of numerically computed quantities by

$$x''(t_{n+1}) \approx \frac{x'_{n+1} - x'_n}{h_n}.$$

Hence, T_{n+1} is approximated by

$$\widehat{T}_{n+1} = \tfrac{1}{2} h_n (x'_{n+1} - x'_n), \tag{11.16}$$

and this is used in (11.12)) to update the step size.[6] We choose a small initial step size $h_0 = 0.1\mathsf{tol}$ and allow the adaptive mechanism to increase it automatically during subsequent steps.

The results are shown in Figure 11.4 for tolerances $\mathsf{tol} = 10^{-2}$ and 10^{-4}. It is not surprising that these show a strong resemblance to those for TS(1) in Figure 11.1. □

Example 11.5

Develop an algorithm based on the trapezoidal rule with automatic step size selection and apply the result to the IVP of Example 11.2.

When h varies with n, the trapezoidal rule becomes

$$x_{n+1} - x_n = \tfrac{1}{2} h_n (f_{n+1} + f_n) \tag{11.17}$$

and the LTE at $t = t_n$ is given by (see Example 4.9)

$$T_{n+1} = -\tfrac{1}{12} h_n^3 x'''(t_n) + \mathcal{O}(h_n^4). \tag{11.18}$$

The step-size-changing formula, (11.4) with $p = 2$, requires an estimate, $\widehat{T}_{n+1}$, of T_{n+1}. This will require us to estimate $x'''(t_n)$.

---

[6]Notice that the values of x'_n and x'_{n+1} are already available, since they are required to apply Euler's method. The cost of calculating $\widehat{T}_{n+1}$ is therefore negligible.

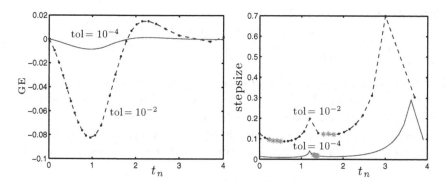

Fig. 11.4 Numerical results for Example 11.4 using Euler's method. Shown are the variation of GE (left) and time step h_n (right) versus time t_n with tolerances tol $= 10^{-2}, 10^{-4}$

This is done by using the Taylor expansions (see Appendix B)

$$x'(t_{n+1}) = x'(t_n) + h_n x''(t_n) + \tfrac{1}{2} h_n^2 x'''(t_n) + \mathcal{O}(h_n^3),$$
$$x'(t_{n-1}) = x'(t_n) - h_{n-1} x''(t_n) + \tfrac{1}{2} h_{n-1}^2 x'''(t_n) + \mathcal{O}(h_{n-1}^3),$$

which make use of $t_{n+1} = t_n + h_n$ and $t_{n-1} = t_n - h_{n-1}$. When they are rearranged as

$$\frac{x'(t_{n+1}) - x'(t_n)}{h_n} = x''(t_n) + \tfrac{1}{2} h_n x'''(t_n) + \mathcal{O}(h_n^2),$$
$$\frac{x'(t_n) - x'(t_{n-1})}{h_{n-1}} = x''(t_n) - \tfrac{1}{2} h_{n-1} x'''(t_n) + \mathcal{O}(h_{n-1}^2),$$

$x''(t_n)$ may be eliminated by subtracting the second of these equations from the first. The resulting expression, when solved for $x'''(t_n)$, leads to

$$x'''(t_n) = \frac{2}{h_n + h_{n-1}} \left(\frac{x'(t_{n+1}) - x'(t_n)}{h_n} - \frac{x'(t_n) - x'(t_{n-1})}{h_{n-1}} \right) + \mathcal{O}(h),$$

where we have used h to denote the larger of h_n and h_{n-1} in the remainder term. Thus, $x'''(t_n)$ can be approximated by

$$x'''(t_n) \approx \frac{2}{h_n + h_{n-1}} \left(\frac{x'_{n+1} - x'_n}{h_n} - \frac{x'_n - x'_{n-1}}{h_{n-1}} \right),$$

since it is written in terms of previously computed quantities x'_{n-1}, x'_n, and x'_{n+1}. By combining these approximations we find that the step size can be updated via

$$h_{\text{new}} = h_n \left| \frac{\text{tol}}{\widehat{T}_{n+1}} \right|^{1/3}, \tag{11.19}$$

where (see equation (11.18))

$$\widehat{T}_{n+1} = -\frac{1}{6}\frac{h_n^3}{h_n + h_{n-1}}\left(\frac{x'_{n+1} - x'_n}{h_n} - \frac{x'_n - x'_{n-1}}{h_{n-1}}\right). \tag{11.20}$$

The integration process is initiated using two steps with very small time steps ($h_0 = h_1 = $ tol, say) and formula (11.19) is then used to predict a suitable step size for subsequent time levels. We see in the right of Figure 11.5 that $h_2 \gg h_1$ and the time steps immediately settle down to an appropriate level. Comparing this figure with Figure 11.2 for TS(2) we see that the step sizes chosen for the two methods are very similar.

For a second-order method the GE is expected to be proportional to $\mathsf{tol}^{2/3}$; this is confirmed in the left part of Figure 11.5 where the scaled global errors $\mathsf{GE}/\mathsf{tol}^{2/3}$ are approximately equal in the two cases.   □

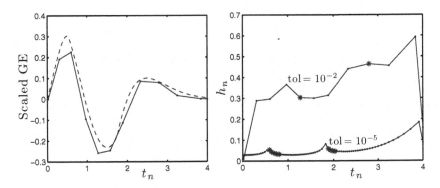

Fig. 11.5 The numerical results for Example 11.5 using the trapezoidal rule. Shown are the scaled GE (that is, $\mathsf{GE}/\mathsf{tol}^{2/3}$) (left) and step sizes h_n (right) versus t for tolerances $\mathsf{tol} = 10^{-2}$ and 10^{-5} (dashed curve). Asterisks ($*$) indicate rejected steps

11.3 Runge–Kutta Methods

We recall (Definition 9.3) that the LTE of an RK method is defined to be the difference between the exact and the numerical solution of the IVP at time $t = t_{n+1}$ under the localizing assumption that $x_n = x(t_n)$; that is, the two quantities were equal at the beginning of the step.

The idea here is to use two (related) RK methods, one of order p and another of order $p+1$, to approximate the LTE. Suppose that these two methods produce solutions $x_{n+1}^{\langle p \rangle}$ and $x_{n+1}^{\langle p+1 \rangle}$. Then the difference $x_{n+1}^{\langle p+1 \rangle} - x_{n+1}^{\langle p \rangle}$ will provide an estimate of the error in the lower order method. Generally, such an approach would be grossly inefficient, since it would involve computing two sets of k-values. This duplication can be avoided by choosing the free parameters in our methods so that the k values for the lower order method are a subset of those of the higher order method (see Exercise 9.13). In this way we get two methods almost for the price of one, and the LTE estimate can be obtained with little additional computation.

Example 11.6 (RK(1,2))

Use Euler's method together with the second-order improved Euler method with Butcher tableau

$$
\begin{array}{c|cc}
0 & & \\
1 & 1 & \\
\hline
& \frac{1}{2} & \frac{1}{2}
\end{array}
$$

to illustrate the construction of a method with adaptive step size control. Use the resulting method to approximate the solution of the IVP of Example 11.2.

Since $k_1 = f(t_n, x_n)$, we observe that $x_{n+1} = x_n + h_n k_1$ is simply the one-stage, first-order method RK(1) (Euler's method). Suppose we denote the result of this calculation by $x^{\langle 1 \rangle}$; that is,

$$x_{n+1}^{\langle 1 \rangle} = x_n + h_n k_1. \tag{11.21}$$

Also, if $x_{n+1}^{\langle 2 \rangle}$ is the result of using the improved Euler method (see Section 9.4), then

$$x_{n+1}^{\langle 2 \rangle} = x_n + \tfrac{1}{2} h_n (k_1 + k_2), \tag{11.22}$$

where $k_2 = f\big(t_n + h_n, x_n + h_n k_1\big)$. The LTE T_{n+1} in Euler's method at $t = t_{n+1}$ is estimated by the difference

$$\widehat{T}_{n+1} = x_{n+1}^{\langle 2 \rangle} - x_{n+1}^{\langle 1 \rangle},$$

so that the updating formula (11.4) gives

$$h_{\text{new}} = h_n \left| \frac{\text{tol}}{\widehat{T}_{n+1}} \right|^{1/2}, \qquad \widehat{T}_{n+1} = \tfrac{1}{2} h_n (k_2 - k_1),$$

and we set $x_{n+1} = x_{n+1}^{\langle 1 \rangle}$. When this method is applied to the IVP (11.9) with $h_0 = \text{tol}$, the scaled GE and the time steps are shown in Figure 11.6 as functions of time for tolerances tol $= 10^{-2}$ and 10^{-4}. These show a strong similarity to the results shown in Figures 11.1 and 11.4 for our other first-order methods. □

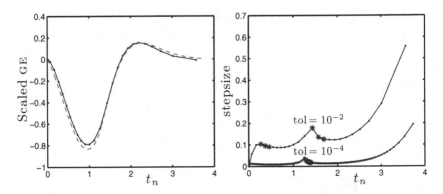

Fig. 11.6 Numerical results for Example 11.2 using RK(1,2). The variation of scaled GE (that is, $GE/tol^{1/2}$) (left) and time step h_n (right) versus time t_n with tolerances $tol = 10^{-2}$ (solid) and 10^{-4} (dashed). Asterisks ($*$) indicate rejected steps

Example 11.7 (RK(2,3))

Use the third-order method with Butcher tableau

$$
\begin{array}{c|cccc}
0 & 0 \\
1 & 1 & 0 \\
\frac{1}{2} & \frac{1}{4} & \frac{1}{4} & 0 \\
\hline
 & \frac{1}{6} & \frac{1}{6} & \frac{2}{3}
\end{array}
$$

to illustrate the construction of a method with adaptive step size control. Use the resulting method to approximate the solution of the IVP of Example 11.2.

When x_{n+1} is computed using only the first two rows of this tableau

$$x_{n+1}^{\langle 2 \rangle} = x_n + \tfrac{1}{2} h_n \big(k_1 + k_2 \big)$$

we find that this method is the second order improved Euler method (see the previous example). On the other hand,

$$x_{n+1}^{\langle 3 \rangle} = x_n + \tfrac{1}{6} h_n \big(k_1 + k_2 + 4 k_3 \big)$$

gives a method which is third-order accurate (see Exercise 11.14). The difference $x_{n+1}^{\langle 3 \rangle} - x_{n+1}^{\langle 2 \rangle}$ leads to the estimate

$$\widehat{T}_{n+1} = \tfrac{1}{3} h_n \big(k_1 + k_2 - 2 k_3 \big)$$

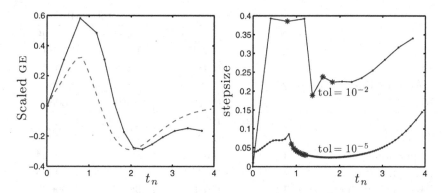

Fig. 11.7 Numerical results for Example 11.2 using RK(2,3). The variation of scaled GE (that is, GE/tol$^{2/3}$) (left) and time step h_n (right) versus time t_n with tolerances tol $= 10^{-2}$ (solid) and 10^{-5} (dashed). Asterisks (∗) indicate rejected steps

for the LTE. Then, with $p = 2$ in (11.4), the update is given by

$$h_{\text{new}} = h_n \left| \frac{\text{tol}}{\widehat{T}_{n+1}} \right|^{1/3}.$$

In our experiments we have used an initial step size $h_0 = \text{tol}$.

When applied to the the IVP of Example 11.2, the results are shown in Figure 11.7. The correlation between the scaled GE with the two tolerances is not as pronounced as with the other methods in this chapter (the correlation improves as tol is reduced). □

11.4 Postscript

A summary of the results obtained with the methods described in this chapter is given in Table 11.2. The main points to observe are that:

(a) methods of the same order have very similar performance;

(b) second-order methods are much more efficient than first-order methods— they deliver greater accuracy with fewer steps.

These conclusions apply only to nonstiff problems. On stiff systems the step size will be restricted by absolute stability and, in these cases, implicit methods would outperform explicit methods independently of their orders. For example, when the IVP of Example 11.3 is solved by the backward Euler method (see

	tol $= 10^{-2}$		tol $= 10^{-4}$	
	No. steps	Max. |GE|	No. steps	Max. |GE|
TS(1)	23(10)	0.081	209(8)	0.0084
Euler	25(7)	0.083	213(6)	0.0084
Backward Euler	24(5)	0.081	213(7)	0.0084
RK(1,2)	27(8)	0.079	216(9)	0.0083
	tol $= 10^{-2}$		tol $= 10^{-5}$	
TS(2)	12(7)	0.023	102(13)	0.00018
trapezoidal	13(2)	0.012	86(11)	0.00014
RK(2,3)	15(4)	0.027	98(11)	0.00015
RK(2,3)(Extrapolated)	14(4)	0.010	98(11)	0.00005

Table 11.2 Summary of numerical results for the methods described in this chapter showing the number of steps to integrate over the interval $(0, 4)$, the number of rejected steps in parentheses, and the maximum $|GE|$ taken over the interval

Exercise 11.8) rather than TS(1), we obtain the results shown in Figure 11.8. On comparing Figures 11.3 and 11.8 we see that for the more relaxed tolerance of 10^{-2} the implicit method is allowed to use time steps that grow smoothly with t because the error control strategy is not detecting any instability.

Rejected time steps impose an additional computational burden; in order to reduce their occurrence, the main formula (11.4) for updating the step size is often altered by including a "safety factor" of 0.9 on the right. The step size used is therefore just 90% of the theoretically predicted value and will lead to roughly 10% more steps being required—thus, this adjustment need only be made if the number of rejected steps (which should be monitored as the integration proceeds) approaches 10% of the number of steps used to date.

In RK methods the choice of step size is based on the estimate of the LTE of the lower order method ($x^{\langle p \rangle}$). However, the solution $x^{\langle p+1 \rangle}$ would normally be expected to give a more accurate approximation. It is common, therefore, to use $x_{n+1} = x_{n+1}^{\langle p+1 \rangle}$ when a step is accepted—the RK process is then said to be in *local extrapolation* mode. The results for RK(2,3) operated in this way are shown in the final row of Table 11.2 and, despite the lack of theoretical justification, we see superior results to the basic RK(2,3) method. The codes for solving IVPs presented in MATLAB, for instance, use local extrapolation. Information on these may be found in the book by Shampine et al. [63]. Moler [53] gives a detailed description of the implementation of a simplified version of a (2,3) Runge–Kutta pair devised by Bogacki and Shampine [4] that is the basis of the MATLAB function ode23.

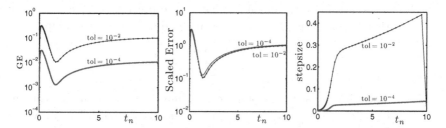

Fig. 11.8 Numerical results for Example 11.3 using the backward Euler method. Shown are the variation of GE (left), the scaled GE (that is, $GE/tol^{1/2}$) (middle), and time step h_n (right) versus time t_n. The tolerances used are tol $= 10^{-2}, 10^{-4}$. These should be compared with the corresponding plots for TS(1) shown in Figure 11.3

EXERCISES

11.1.* Show that the conclusion from Example 11.1 is true regardless of the accuracy requested. That is, if h_0 is chosen so that $|x(t_1) - x_1| = $ tol, then $|x(t_2) - x_2| = $ tol is unachievable for any choice of $h_1 > 0$.

11.2.* Repeat Example 11.1 for the backward Euler method

$$x_{n+1} = x_n + 2h_n t_{n+1}, \qquad t_{n+1} = t_n + h_n.$$

(Note: in this case we require $|x(t_n) - x_n| = 0.01$ for each n.)

11.3.* Show that the TS(1) method described in Example 11.2 applied to the IVP $x'(t) = \lambda x(t)$, $x(0) = 1$, leads to

$$x_{n+1} = (1 + h_n \lambda)x_n, \qquad h_{new} = \left| \frac{2\,tol}{\lambda^2 x_n} \right|^{1/2}, \qquad x_0 = 1.$$

[This exercise is explored further in Exercise 13.12.]

11.4.* Prove for the TS(1) method described in Example 11.2 that $|x_{n+1}| < |x_n|$ (the analogue of absolute stability in this case) for $t_n > \frac{1}{2}$ if

$$h_n < \frac{2}{2t_n - 1}.$$

This implies that the step size must tend to zero as $t \to \infty$ despite the fact that the solution tends to zero. The final steps in the experiments reported in Figure 11.1 (right) were rejected so as to avoid h_n exceeding this value. (This is a situation where using an implicit LMM would be advantageous.)

11.5.** Use a similar argument to that in the previous exercise to prove that $|x_{n+1}| < |x_n|$ for the TS(2) method described in Example 11.2 if

$$h_n < \frac{2(2t_n - 1)}{(2t_n - 1)^2 - 2}$$

when $t_n > \frac{1}{2} + \sqrt{\frac{2}{3}}$. Compare the limit imposed on h_n by this inequality when $t_n \approx 3$ with the two final rejected steps shown in Figure 11.2 for tol $= 10^{-2}$.

11.6.* Show that

$$x'''(t) = (1 - 2t)\big[(1 - 2t)^2 - 6\big]x(t)$$

for the IVP (11.9). Use this in conjunction with Equation (11.7) for $p = 2$ to explain the local maxima in the plot of h_n versus t in Figure 11.2 for $t \leq 2$.

11.7.** Devise a step-sizing algorithm for TS(3) applied to the IVP (11.9).

11.8.* Explain why the step-size-changing formula for the backward Euler method is identical to that for Euler's method described in Example 11.4.

11.9.** Show that the negative of the expression given by (11.16) provides an estimate for the LTE of the backward Euler method. Show that this estimate may also be derived via Milne's device (8.11) for the predictor-corrector pair described in Example 8.2.

11.10.** Calculate x_1, t_1, and h_1 when the IVP (11.9) is solved with $h_0 = $ tol, tol $= 10^{-2}$, and (a) TS(1), (b) TS(2) and (c) Euler's method (as in Example 11.4).

11.11.** Calculate (t_1, x_1) and (t_2, x_2) when the IVP (11.9) is solved using the trapezoidal rule with $h_0 = h_1 = $ tol, tol $= 10^{-2}$ (see Example 11.5).

11.12.** The coefficients of LMMs with step number $k > 1$ have to be adjusted when the step sizes used are not the same for every step. There is more than one way to do this, but for the AB(2) method (Section 4.1.2) a popular choice is

$$x_{n+1} = x_n + \tfrac{1}{2}h_n\left[\left(2 + \frac{h_n}{h_{n-1}}\right)x'_n - \frac{h_n}{h_{n-1}}x'_{n-1}\right]. \qquad (11.23)$$

The LTE of this method at $t = t_n$ is defined by

$$\mathscr{L}_h x(t_n) = x(t_{n+1}) - x(t_n) - \tfrac{1}{2}h_n\left[\left(2 + \frac{h_n}{h_{n-1}}\right)x'(t_n) - \frac{h_n}{h_{n-1}}x'(t_{n-1})\right]. \qquad (11.24)$$

Show, by Taylor expansion, that

$$\mathscr{L}_h x(t_n) = \tfrac{1}{12}(2h_n + 3h_{n-1})h_n^2 x'''(t_n) + \mathcal{O}(h^4),$$

where $h = \max\{h_{n-1}, h_n\}$.

11.13.*** Suppose that the Adams–Bashforth method AB(2) and the trapezoidal rule are to be used in predictor-corrector mode with variable
step sizes. Let $x_{n+1}^{[0]}$ and x_{n+1} denote, respectively, the values computed at $t = t_{n+1}$ by Equations (11.23) and (11.17). By following a
process similar to that described in Exercise 8.18 for constant step
sizes and using the expression for the LTE of the AB(2) method given
in the previous exercise, show that the generalization of Milne's device (8.11) to this situation gives the estimate

$$\widehat{T}_{n+1} = -\frac{1}{3}\frac{h_n}{h_n + h_{n-1}}\left(x_{n+1} - x_{n+1}^{[0]}\right)$$

for the LTE T_{n+1}. Verify that this agrees with the estimate (11.20)
derived earlier in this chapter[7]. Verify that the result agrees with
that in Exercise 8.12 when $h_{n-1} = h_n = h$.

11.14.* Show, by using the results of Table 9.6, or otherwise, that the RK
method defined by the tableau in Example 11.7 is consistent of order 3.

11.15.*** Show that the modified Euler method (Table 9.5 with $\theta = 1$) uses
two of the same k-values as Kutta's third-order method (Table 9.7).

Write down an adaptive time-step process based on this pair of methods and verify, when it is applied to the IVP (11.9) with $h_0 = $ tol
and tol $= 0.01$, that $x_2 = 1.252$ and $t_2 = 0.321$.

---

[7]Writing the estimated LTE for the trapezoidal rule in terms of the difference
between predicted and corrected values is more common than using the form (11.20).

12
Long-Term Dynamics

There are many applications where one is concerned with the long-term behaviour of nonlinear ODEs. It is therefore of great interest to know whether
this behaviour is accurately captured when they are solved by numerical methods. We will restrict attention to *autonomous* systems of the form[1]

$$\boldsymbol{x}'(t) = \boldsymbol{f}(\boldsymbol{x}(t)) \tag{12.1}$$

in which $\boldsymbol{x} \in \mathbb{R}^m$ and the function $\boldsymbol{f}$ depends only on $\boldsymbol{x}$ and not explicitly on t.
Equations of this type occur sufficiently often to make their study useful and
their analysis is much easier than for their non-autonomous counterparts where
$\boldsymbol{f}$ depends also on t. One immediate advantage is that we can work in the phase
plane. For example, if $\boldsymbol{x} = [x, y]^{\mathrm{T}} \in \mathbb{R}^2$, we may investigate the evolution of the
solution as a curve $(x(t), y(t))$ parameterized by t. See, for example, Figure 1.3
(right) for describing solutions of the Lotka–Volterra equations.

   We will first discuss the behaviour of solutions of the differential equations—
the continuous case. This will be followed by an application of similar principles
to numerical methods applied to the differential equations—the discrete case.
The aim is to deduce qualitative information regarding solutions—building up
a picture of the behaviour of solutions without needing to find any general
solutions (which is rarely possible). The approach is to look for constant solutions and then to investigate, by linearization, how nearby solutions behave.

---

[1]Of course, as described in Section 1.1, any non-autonomous ODE may be trans-
formed into autonomous form. However, this chapter focuses on autonomous ODEs
with fixed points, which rules out such a transformation.

D.F. Griffiths, D.J. Higham, *Numerical Methods for Ordinary Differential Equations*,
Springer Undergraduate Mathematics Series, DOI 10.1007/978-0-85729-148-6_12,
© Springer-Verlag London Limited 2010

Our treatment will necessarily be brief; for further details we recommend the books by Braun [5, Chapter 4], Stuart and Humphries [65] and Verhulst [68].

12.1 The Continuous Case

Suppose that $x(t) = x^*$, where x^* is a constant, is a solution of (12.1). Clearly we must have $f(x^*) = 0$, and this motivates a definition.

Definition 12.1 (Fixed Point)

If $x^* \in \mathbb{R}^m$ satisfies $f(x^*) = 0$ then x^* is called a *fixed point*[2] of the system (12.1).

Suppose that a solution becomes close to a fixed point. Will this solution be attracted towards it or be repelled away from it? The following definition allows us to phrase this mathematically.

Definition 12.2 (Linear Stability)

A fixed point x^* of (12.1) is *linearly stable* (or locally attracting) if there exists a neighbourhood[3] around x^* such that any solution $x(t)$ entering this neighbourhood satisfies $x(t) \to x^*$ as $t \to \infty$.

To investigate linear stability we write a solution of the ODE system in the form

$$x(t) = x^* + \varepsilon u(t), \qquad (12.2)$$

where ε is a "small" real number, and ask whether $u(t)$ grows or decays. We do this by substituting (12.2) into (12.1) to obtain

$$\varepsilon u'(t) = f(x^* + \varepsilon u(t)). \qquad (12.3)$$

Using the Taylor expansion of a function of several variables (see Appendix C)

$$f(x^* + \varepsilon u(t)) = f(x^*) + \varepsilon \frac{\partial f}{\partial x}(x^*)u(t) + \mathcal{O}(\varepsilon^2)$$

---

[2] Also known as an equilibrium point, critical point, rest state, or steady state.

[3] The phrase "neighbourhood around x^*" means the set of points z within a sphere of radius δ centred at x^*, for any positive value of δ. Thus, neighbourhoods can be arbitrarily small but must contain x^* strictly in their interior.

and neglecting the $\mathcal{O}(\varepsilon^2)$ term, (12.3) becomes, to first order,

$$\boldsymbol{u}'(t) = A\boldsymbol{u}(t), \tag{12.4}$$

where A is the $m \times m$ matrix

$$A = \frac{\partial \boldsymbol{f}}{\partial \boldsymbol{x}}(\boldsymbol{x}^*), \tag{12.5}$$

known as the Jacobian of the system (12.1) at $\boldsymbol{x}^*$. In two dimensions, where $\boldsymbol{x} = [x, y]^{\mathrm{T}} \in \mathbb{R}^2$ and $\boldsymbol{f} = [f, g]^{\mathrm{T}}$, the matrix A has the 2×2 form

$$A = \begin{bmatrix} f_x & f_y \\ g_x & g_y \end{bmatrix}.$$

The system (12.4) is known as the linearization of (12.1) at $\boldsymbol{x}^*$, the idea being that in the neighbourhood of $\boldsymbol{x}^*$ the behaviour of solutions of (12.1) can be deduced by studying solutions of the linear system (12.4). From (12.2) we see that $\boldsymbol{u}(t)$ shows us how a small perturbation from $\boldsymbol{x}^*$ evolves over time. A sufficient condition that the solution $\boldsymbol{x} = \boldsymbol{x}^*$ be linearly stable is that $\boldsymbol{u}(t) \to 0$ in (12.4) as $t \to \infty$. The next result follows from Theorem 7.3.

Theorem 12.3 (Linear Stability)

A fixed point $\boldsymbol{x}^*$ of the system $\boldsymbol{x}'(t) = \boldsymbol{f}(\boldsymbol{x}(t))$ is linearly stable if $\Re(\lambda_A) < 0$ for every eigenvalue λ_A of the Jacobian $A = \dfrac{\partial \boldsymbol{f}}{\partial \boldsymbol{x}}(\boldsymbol{x}^*)$.

Example 12.4

Determine the fixed points of the logistic equation $x'(t) = 2x(t)\big(1 - x(t)\big)$ and investigate whether they are linearly stable (see Example 2.2).

In this example we are dealing with scalar quantities ($m = 1$) with $f(x) = 2x(1 - x)$. The fixed points are given by $f(x) = 0$, and so $x_1^* = 0$ and $x_2^* = 1$.
   The Jacobian is also scalar. It is given by the 1×1 matrix

$$\frac{\partial f}{\partial x}(x) = 2 - 4x,$$

so that

$$\frac{\partial f}{\partial x}(0) = 2 > 0 \quad \text{and} \quad \frac{\partial f}{\partial x}(1) = -2 < 0.$$

Hence, the fixed point $x_1^* = 0$ is locally repelling—solutions starting close to $x = 0$ will move further away—while $x_2^* = 1$ will attract nearby solutions. This is illustrated in Figure 2.3, where a solution starting from $x(0) = 0.2$ moves away from $x = 0$ and is attracted to $x = 1$ as $t \to \infty$. $\qquad\square$

Example 12.5

Investigate the nature of the fixed points of the Lotka–Volterra system (see Example 1.3[4])

$$x'(t) = 0.05x(t)\big(1 - 0.01y(t)\big),$$
$$y'(t) = 0.1y(t)\big(0.005x(t) - 2\big). \tag{12.6}$$

Now

$$\boldsymbol{f}(\boldsymbol{x}) = \begin{bmatrix} f(x,y) \\ g(x,y) \end{bmatrix}, \qquad f(x,y) = 0.05x(1 - 0.01y), \quad g(x,y) = 0.1y(0.005x - 2).$$

The fixed points are solutions of the simultaneous nonlinear algebraic equations $f(x,y) = 0$ and $g(x,y) = 0$. These lead to the two fixed points

$$\boldsymbol{x}_1^* = \begin{bmatrix} 0 \\ 0 \end{bmatrix} \quad \text{and} \quad \boldsymbol{x}_2^* = \begin{bmatrix} 400 \\ 100 \end{bmatrix}.$$

The Jacobian of the system is found to be

$$\frac{\partial \boldsymbol{f}}{\partial \boldsymbol{x}}(\boldsymbol{x}) = \begin{bmatrix} 0.05(1 - 0.01y) & -0.0005x \\ 0.0005y & 0.1(0.005x - 2) \end{bmatrix}.$$

At the first fixed point $\boldsymbol{x}_1^*$ the Jacobian is

$$\frac{\partial \boldsymbol{f}}{\partial \boldsymbol{x}}(\boldsymbol{x}_1^*) = \begin{bmatrix} 0.05 & 0 \\ 0 & -0.2 \end{bmatrix},$$

whose eigenvalues are $\lambda_1 = 0.05$ and $\lambda_2 = -0.2$. One of these is positive, so the origin is not an attracting fixed point. At the second fixed point $\boldsymbol{x}_2^*$ the Jacobian is

$$\frac{\partial \boldsymbol{f}}{\partial \boldsymbol{x}}(\boldsymbol{x}_2^*) = \begin{bmatrix} 0 & -0.2 \\ 0.05 & 0 \end{bmatrix},$$

whose eigenvalues are $\lambda_1 = \pm 0.1i$. The real parts of both eigenvalues are zero, so it is not possible to deduce the precise behaviour of the original system without taking the nonlinear terms into account. This would take us beyond the scope of this book (see Braun [5, Section 4.10] or Verhulst [68] for a detailed study); we observe from Figure 1.3 that the motion is periodic (indicated by the closed curves) around $(400, 100)$, and this is in keeping with the findings of Section 7.3, where imaginary eigenvalues were seen to be associated with oscillatory behaviour.   □

---

[4]We have used dependent variables x, y here so that u, v can be used for the linearized system.

12.2 The Discrete Case

In this section we study the dynamical behaviour of the discrete map

$$\boldsymbol{x}_{n+1} = \boldsymbol{F}(\boldsymbol{x}_n), \qquad (12.7)$$

where $\boldsymbol{x}_n \in \mathbb{R}^m$ and $\boldsymbol{F}(\boldsymbol{x})$ is assumed to be a continuously differentiable function of the m-dimensional vector $\boldsymbol{x}$. In the context of numerical methods for solving systems of ODEs such as (12.1), $\boldsymbol{F}$ is parameterized by the step size h: Euler's method, for example, leads to

$$\boldsymbol{F}(\boldsymbol{x}) = \boldsymbol{x} + h\boldsymbol{f}(\boldsymbol{x}). \qquad (12.8)$$

Definition 12.6 (Fixed Point)

If $\boldsymbol{x}^* \in \mathbb{R}^m$ satisfies $\boldsymbol{x}^* = \boldsymbol{F}(\boldsymbol{x}^*)$ then $\boldsymbol{x}^*$ is called a *fixed point* of the discrete map (12.7).

Definition 12.7 (Linear Stability)

A fixed point $\boldsymbol{x}^*$ of (12.7) is *linearly stable* (or locally attracting) if there exists a neighbourhood around $\boldsymbol{x}^*$ such that any solution $\boldsymbol{x}_n$ entering this neighbourhood satisfies $\boldsymbol{x}_n \to \boldsymbol{x}^*$ as $n \to \infty$.

Mimicking the continuous case, we can investigate linear stability by writing

$$\boldsymbol{x}_n = \boldsymbol{x}^* + \varepsilon \boldsymbol{u}_n.$$

Substituting this into (12.7) and using the Taylor expansion

$$\boldsymbol{F}(\boldsymbol{x}^* + \varepsilon \boldsymbol{u}_n) = \boldsymbol{F}(\boldsymbol{x}^*) + \varepsilon \frac{\partial \boldsymbol{F}}{\partial \boldsymbol{x}} \boldsymbol{u}_n + \mathcal{O}(\varepsilon^2)$$

we obtain, on neglecting terms of order $\mathcal{O}(\varepsilon^2)$, the linearization of (12.7) at $\boldsymbol{x}^*$:

$$\boldsymbol{u}_{n+1} = B\boldsymbol{u}_n, \qquad (12.9)$$

where the $m \times m$ matrix

$$B = \frac{\partial \boldsymbol{F}}{\partial \boldsymbol{x}}(\boldsymbol{x}^*)$$

is the Jacobian of the function $\boldsymbol{F}(\boldsymbol{x})$ at $\boldsymbol{x}^*$. The following is an analogue of Theorem 12.3.

Theorem 12.8 (Linear Stability)

A fixed point x^* of the system $x_{n+1} = F(x_n)$ is linearly stable if $|\lambda_B| < 1$ for every eigenvalue λ_B of the Jacobian $B = \dfrac{\partial F}{\partial x}(x^*)$.

Proof

See, for example, Kelley [41, Theorem 1.3.2 and Chapter 4]. □

Some insight into Theorem 12.8 can be gleaned (using an argument similar to that used in Section 8.2) by observing that if λ_B is an eigenvalue of B with corresponding eigenvector v then, choosing $u_0 = v$, the solution of (12.9) is $u_n = \lambda_B^n v$. It follows that this u_n cannot tend to zero as $n \to \infty$ when $|\lambda_B| \geq 1$.

The next theorem gives an indication of the relationship between the fixed points of the ODE and those that result when the ODE is solved by Euler's method. Its conclusions remain true for all LMMs, but we will not prove this.

Theorem 12.9

Suppose that Euler's method is applied to the ODE system (12.1) leading to the discrete map (12.7) with $F(x)$ defined in (12.8). Then x^* is a fixed point of (12.7) if, and only if, it is a fixed point of (12.1).

Suppose that x^* is a linearly stable fixed point of (12.1) with $\Re(\lambda_A) < 0$ for every eigenvalue λ_A of the Jacobian $A = \dfrac{\partial f}{\partial x}(x^*)$. Then it is also a linearly stable fixed point of (12.7) provided that $h\lambda_A \in \mathcal{R}$, the region of absolute stability of Euler's method (Figure 6.6), for every eigenvalue λ_A of A.

Proof

The proof of the first part is left as Exercise 12.2. For the second part, $F(x)$ is defined by (12.8) so its Jacobian is given by

$$\frac{\partial F}{\partial x}(x) = I + h\frac{\partial f}{\partial x}(x),$$

where I is the $m \times m$ identity matrix. Therefore, the matrix B in (12.9) is related to the matrix A in (12.4) through

$$B = I + hA,$$

where I is the $m \times m$ identity matrix, and it follows that the eigenvalues λ_A of A and λ_B of B are related by (see Exercise 12.3)

$$\lambda_B = 1 + h\lambda_A.$$

We are assuming that x^* is a linearly stable fixed point of (12.1) and the eigenvalues of A satisfy $\Re(\lambda_A) < 0$. It follows from Theorem 12.8 that x^* will also be a linearly stable fixed point of Euler's method provided $|1 + h\lambda_A| < 1$, which is the condition derived in Example 6.7 for $\widehat{h} = h\lambda_A$ to lie in the region of absolute stability. $\square$

We will show in Example 12.11 that the first part of the previous theorem is not true for the modified Euler method. In fact it does not hold for general explicit RK methods, as they may admit fixed points that are not fixed points of the ODE system.

Example 12.10

Investigate the linear stability of the fixed points of the map obtained when Euler's method is applied to the ODE $x'(t) = 2x(t)\big(1 - x(t)\big)$ that was analysed in Example 12.4.

The map is given by

$$x_{n+1} = F(x_n), \quad \text{where } F(x) = x + hf(x), \tag{12.10}$$

and $f(x) = 2x(1 - x)$. It is readily shown that the fixed points are $x^* = 0, 1$. The Jacobian is the 1×1 matrix

$$\frac{\partial F}{\partial x}(x) = 1 + 2h(1 - 2x)$$

and so

$$\frac{\partial F}{\partial x}(0) = 1 + 2h \quad \text{and} \quad \frac{\partial F}{\partial x}(1) = 1 - 2h.$$

Since $\frac{\partial F}{\partial x}(0) > 1$ for all $h > 0$, the fixed point $x^* = 0$ is not linearly stable and all solutions close to $x_n = 0$ are repelled.

At the other fixed point $x^* = 1$ we have

$$\left| \frac{\partial F}{\partial x}(1) \right| = |1 - 2h| < 1$$

if, and only if, $0 < h < 1$, in which case it is linearly stable.

In this example $\lambda_A = f_x(1) = -2$ and the interval of absolute stability of Euler's method is $h\lambda_A \in (-2, 0)$. This translates to $0 < h < 1$, confirming the relationship between linear stability of true fixed points and absolute stability of the numerical method identified in Theorem 12.9.

The results of an experiment to test these conclusions are shown in Figure 12.1. Here, we carried out the iteration (12.10) for 500 steps and then plotted the points (h, x_n) for the next 20 steps (where, if it possesses one, the

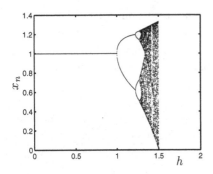

Fig. 12.1 Bifurcation diagram for Euler's method applied to the logistic equation of Example 12.4. The points (h, x_n), $500 < n \le 520$ are plotted for each h.

sequence would be expected to have reached its stable steady state to within graphical accuracy). This process was repeated for many values of h between 0 and 2 and several initial values were used for each value of h.

We see in Figure 12.1 that the sequence x_n seems to settle down to the fixed point $x^* = 1$ for $0 < h < 1$. The "fixed point" branches to the right of the value $h = 1, x = 1$—this is a "period 2" solution, where alternate values of x_n take the values $a + b$ and $a - b$ (see Exercise 12.4). This branching is known as a *bifurcation* (literally to divide into two branches).

The period 2 solution can be found analytically by studying the fixed points of the iterated map $x_{n+1} = \mathcal{F}(x_n)$, where $\mathcal{F}(x) = F(F(x))$, and its linear stability analysed by calculating the Jacobian $\partial \mathcal{F} / \partial x$ at the fixed points. In this way, it can be shown that the period 2 solution loses stability at $h = \frac{1}{2}\sqrt{6} \approx 1.22$, at which a further period-doubling occurs leading to a period 4 solution. These period-doublings continue along with other high-period solutions as h increases. For certain values of h, the solution becomes "chaotic"—defined loosely as a sequence that does not repeat itself as $n \to \infty$.

The bifurcation to a period 2 solution occurs when h increases so that $h\lambda_A$ leaves the region of absolute stability. This behaviour contrasts with that when absolute stability is lost in linear systems, where $|x_n| \to \infty$ as $n \to \infty$ (see, for example, Figure 6.2).

For detailed studies of similar quadratic maps see Thompson and Stewart [66, Section 9.2].   □

Our next example contains features common to many RK methods but not present in the previous example.

Example 12.11

Determine the fixed points of the modified Euler method described in Example 9.1 when applied to the scalar equation $x'(t) = 2x(t)\big(1 - x(t)\big)$ and examine their linear stability.

The method is defined by the recurrence

$$x_{n+1} = x_n + hf\left(x_n + \tfrac{1}{2}hf(x_n)\right),$$

where $f(x) = 2x(1-x)$. To find the fixed points we set $x_{n+1} = x_n = x$ so that they satisfy the equation[5]

$$f\left(x + \tfrac{1}{2}hf(x)\right) = 0.$$

Since $f(y) = 0$ has the two roots $y = 0, 1$, the fixed points satisfy

$$x + \tfrac{1}{2}hf(x) = 0 \text{ or } 1. \tag{12.11}$$

These equations then lead to the four fixed points (see Exercise 12.5)

$$x_1^* = 0, \quad x_2^* = 1, \quad x_3^* = 1/h, \quad \text{and} \quad x_4^* = 1 + 1/h. \tag{12.12}$$

With $F(x) = x + hf(x + \tfrac{1}{2}hf(x))$ the Jacobian is given by

$$F'(x) = 1 + hf'\left(x + \tfrac{1}{2}hf(x)\right)\left(1 + \tfrac{1}{2}hf'(x)\right) \tag{12.13}$$

with which it may be verified that

1. $x_1^* = 0$ is not linearly stable for any $h > 0$;

2. $x_2^* = 1$ is linearly stable for $0 < h < 1$;

3. $x_3^* = 1/h$ is linearly stable for $1 < h < \tfrac{1}{2}(1 + \sqrt{5}) \approx 1.62$;

4. $x_4^* = 1 + 1/h$ is linearly stable for $1 < h < \tfrac{1}{2}(-1 + \sqrt{5}) \approx 0.62$.

These results are confirmed by the bifurcation diagram shown in Figure 12.2 (left). There is some cause for some concern, since the numerical method has linearly stable fixed points (x_3^* and x_4^*) that are not fixed points of the ODE—they are so-called *spurious fixed points*—and so numerical experiments could lead to false conclusions being drawn regarding the dynamical properties of the system being simulated. However, since these spurious fixed points depend on h they may be detected by repeating the simulation with a different value of h—any appreciable change to the fixed point would signal that it is spurious. The results of such an experiment are shown in Figure 12.2 (right) where the solution reaches the fixed point $x_3^* = 1/h = 0.8$ when $h = 1.25$, but when h is reduced to 0.625 it approaches the correct fixed point $x_1^* = 1$.

When the fixed points x_2^* and x_1^* lose stability as h is increased there appears to be a sequence of period-doubling bifurcations similar to those observed in the previous example—the dynamics become too complicated for us to summarize here. The occurrence of spurious fixed points is not restricted to the modified Euler method—it is common to all explicit RK methods. A more detailed investigation of the dynamical behaviour of RK methods is presented in Griffiths et al. [25]. □

---

[5]When $f(x)$ is a polynomial of degree d in x, then $f(x + \tfrac{1}{2}hf(x))$ will be a polynomial of degree d^2.

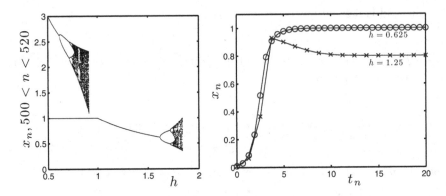

Fig. 12.2 A bifurcation diagram for the modified Euler method applied to the logistic equation of Example 12.4 is shown on the left. On the right are individual solutions from the method with $h = 1.25\,(\times)$ and $h = 0.625\,(\circ)$

EXERCISES

12.1.** Show that the system

$$\begin{aligned} x'(t) &= y(t) - y^2(t), \\ y'(t) &= x(t) - x^2(t) \end{aligned} \tag{12.14}$$

has four fixed points and investigate their linear stability. (A phase portrait of solutions to this system is shown in Figure 13.4 (right).)

12.2.* Prove the first part of Theorem 12.9. That is, prove that $f(x^*) = 0$ implies that $F(x^*) = 0$ and vice versa when f and F are related via (12.8).

12.3.* If $m \times m$ matrices A and B are related via $B = I + hA$, prove that they share the same eigenvectors and that the corresponding eigenvalues are related via $\lambda_B = 1 + h\lambda_A$.

12.4.** Verify, by substitution, that the difference equation (12.10) has a solution of the form $x_n = a + (-1)^n b$, where

$$a = \frac{1 + h}{2h} \quad \text{and} \quad b = \frac{\sqrt{h^2 - 1}}{2h},$$

which is real for $h > 1$.

12.5.** Verify the expressions (12.12) for the fixed points and (12.13) for the Jacobian in Example 12.11.

12.6.** Repeat the calculations of fixed points and their linear stability in Example 12.11 for the function $f(x) = \alpha x(1 - x)$, where α is a positive constant. Check your results with those given in the example when $\alpha = 2$.

12.7.*** Consider the AB(2) method (4.10) applied to the scalar equation $x'(t) = f(x(t))$. Show that this may be written in the form of the map (12.7) by defining $y_n = x_{n+1}$, $z_n = x_n$ (which implies $z_{n+1} = y_n$)

$$\boldsymbol{x}_n = \begin{bmatrix} y_n \\ z_n \end{bmatrix}, \quad \text{and} \quad \boldsymbol{F}(\boldsymbol{x}_n) = \begin{bmatrix} y_n + \frac{1}{2}h\big(3f(y_n) - f(z_n)\big) \\ y_n \end{bmatrix}.$$

In the case $f(x) = 2x(1 - x)$, show that there are two fixed points $\boldsymbol{x}^* = [0, 0]^{\mathrm{T}}$ and $\boldsymbol{x}^* = [1, 1]^{\mathrm{T}}$.

Verify that the Jacobian of the map is given by

$$\frac{\partial \boldsymbol{F}}{\partial \boldsymbol{x}}(\boldsymbol{x}) = \begin{bmatrix} 1 + \frac{3}{2}hf'(y) & -\frac{1}{2}hf'(z) \\ 1 & 0 \end{bmatrix}$$

and, hence, investigate the linear stability of the fixed points. Do your results agree with the bifurcation diagram shown in Figure 12.3?

12.8.** Suppose the separable Hamiltonian problem (see (15.14))

$$p'(t) = -V'(q),$$
$$q'(t) = \quad T'(p)$$

is such that $V'(0) = T'(0) = 0$, $V''(0) > 0$ and $T''(0) > 0$. Show that the Jacobian at the equilibrium point $p = q = 0$ has purely imaginary eigenvalues.

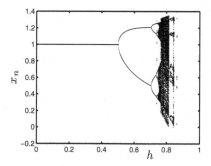

Fig. 12.3 Bifurcation diagram for the AB(2) method applied to the logistic equation of Exercise 12.7

12.9.*** Consider the application of the trapezoidal rule to the logistic
equation

$$x'(t) = x(t)(X - x(t)),$$

in which X is a positive constant. The numerical solution is studied
in the $x_n x_{n+1}$ phase plane—that is, we study the set of points having
coordinates (x_n, x_{n+1}).

(a) Show that the points (x_n, x_{n+1}) lie on a circle passing through
the origin having centre at

$$\left(\frac{X}{2} + \frac{1}{h}, \frac{X}{2} - \frac{1}{h} \right).$$

(b) Explain why the fixed points of the numerical method should lie
at the intersection of the circle with the line $x_{n+1} = x_n$.

(c) Calculate the Jacobian dx_{n+1}/dx_n of the mapping and deduce
that one fixed point is stable while the other is unstable for all
$h > 0$.

Show also that the Jacobian evaluated at either fixed point is
positive for $hX < 2$.

(d) With $x_0 = 1$, verify the values of x_1 given in the table below for
$hX = 1$ (to represent $0 < hX < 2$) and for $hX = 5$ (to represent
$hX > 2$) with $X = 10$.

n	0	1	2	3	4	5
$hX = 5$	1.000	7.690	10.584	9.720	10.114	9.950
$hX = 1$	1.000	2.349	4.484	6.807	8.523	9.424

Sketch the situation described in parts (a) and (b) for $hX = 1$
and for $hX = 5$ with $X = 10$ and use the data given in the table
to draw a cobweb diagram[6] in each case.

Observe that x_n approaches the fixed point monotonically when
$hX < 2$, but not when $hX > 2$.

---

[6]The figure created by joining $(0, x_0)$ to (x_0, x_1) then this point to (x_1, x_1), which
is joined to (x_1, x_2), and so forth is known as a cobweb diagram.

13
Modified Equations

13.1 Introduction

Thus far the emphasis in this book has been focused firmly on the solutions of IVPs and how well these are approximated by a variety of numerical methods. This attention is now shifted to the numerical method (primarily LMMs) and we ask whether the numerically computed values might be closer to the solution of a modified differential equation than they are to the solution of the original differential equation. At first sight this may appear to introduce an unnecessary level of complication, but we will see in this chapter (as well as those that follow on geometric integration) that *constructing a new ODE that very accurately approximates the numerical method* can provide important insights about our computations.

If we suppose that the original IVP has solution $x(t)$ and the numerical solution is x_n at time t_n, we then look for a new function $y(t)$, the solution of a nearby ODE, such that $y(t_n)$ is closer than $x(t_n)$ to x_n at time t_n. Since $e_n = x(t_n) - x_n \approx x(t_n) - y(t_n)$, properties that may be deduced concerning $y(t)$ can be translated into properties of x_n and the difference between the curves $x(t)$ and $y(t)$ will give an idea of the global error. The differential equation satisfied by $y(t)$ is called a *modified equation.*

The most common way of deriving such an equation is to show that the method being studied has a higher order of consistency to the modified equation than it does to the original ODE. This is the approach that we will adopt. We will also show that modified equations are not unique, each numerical method

D.F. Griffiths, D.J. Higham, *Numerical Methods for Ordinary Differential Equations*,
Springer Undergraduate Mathematics Series, DOI 10.1007/978-0-85729-148-6_13,
© Springer-Verlag London Limited 2010

has an unlimited number of modified equations of any given order of accuracy—
we generally choose the simplest equation subject to the requirement that $y(t)$
should be a smooth function.

13.2 One-Step Methods

In this section we shall construct modified equations for Euler's method and
a number of variants that are more appropriate for different types of ODE
systems.

Example 13.1 (Euler's Method)

Determine a modified equation corresponding to Euler's method applied to the
autonomous IVP $x'(t) = f(x(t))$, $x(0) = \eta$.

The numerical method is, in this case, $x_{n+1} = x_n + hf_n$, with $x_0 = \eta$. Regarding
this as a one-stage RK we recall that, as in Chapter 9, the LTE of the method
may be computed from

$$T_{n+1} = x(t_{n+1}) - x_{n+1} \tag{13.1}$$

under the localizing assumption that $x_n = x(t_n)$ (see Definition 9.3). It was
shown in Section 9.3 that $T_{n+1} = \mathcal{O}(h^2)$.

We now construct a modified equation—or more correctly a modified IVP—
of the form

$$y'(t) = f(y(t)) + h^p g(y(t)), \qquad t > t_0, \tag{13.2}$$

with $y(t_0) = \eta$. The integer p and function $g(y)$ are determined so that the
LTE, calculated from[1]

$$\widehat{T}_{n+1} = y(t_{n+1}) - x_{n+1}, \tag{13.3}$$

is of higher order in h than the standard quantity (13.1) when the localizing
assumption that $x_n = y(t_n)$ is employed.

We shall require the Taylor expansion

$$\begin{aligned}
y(t+h) &= y(t) + hy'(t) + \tfrac{1}{2}h^2 y''(t) + \mathcal{O}(h^3) \\
&= y(t) + h\big(f(y(t)) + h^p g(y(t))\big) \\
&\quad + \tfrac{1}{2}h^2 \left(\frac{\mathrm{d}f}{\mathrm{d}y}(y(t))f(y(t)) + h^p \frac{\mathrm{d}g}{\mathrm{d}y}(y(t))g(y(t)) \right) + \mathcal{O}(h^3),
\end{aligned}$$

---

[1] We use a circumflex on the LTE to distinguish it from the standard definition
based on $x(t)$.

where we have taken the time derivative of the right-hand side of (13.2) using the chain rule in order to find $y''(t)$. If we now choose $p = 1$ and reorganize the terms, we find

$$y(t + h) = y(t) + hf(y(t)) + h^2\left(g(y(t)) + \frac{1}{2}\frac{df}{dy}(y(t))f(y(t))\right) + \mathcal{O}(h^3). \quad (13.4)$$

It follows from the localizing assumption $x_n = y(t_n)$ that $x_{n+1} = y(t_n) + hf(y(t_n))$ and so, from (13.3),

$$\widehat{T}_{n+1} = h^2\left(g(y(t_n)) + \frac{1}{2}\frac{df}{dy}(y(t_n))\right) + \mathcal{O}(h^3).$$

Hence, by choosing

$$g(y(t)) = -\frac{1}{2}\frac{df}{dy}(y(t))f(y(t))$$

we shall have $\widehat{T}_{n+1} = \mathcal{O}(h^3)$ so, while the method $x_{n+1} = x_n + hf_n$ is a *first-order* approximation to $x'(t) = f(x(t))$, it is a *second-order* approximation of

$$y'(t) = \left(1 - \frac{1}{2}h\frac{df}{dy}(y(t))\right)f(y(t)). \quad (13.5)$$

This is our modified equation: it is the original ODE modified by the addition of a small $\mathcal{O}(h)$ term—the order of the additional term being, generally, the same as the order of accuracy of the method. We deduce from (13.5) that $|y'(t)| < |f(y(t))|$ when $df/dy > 0$. Thus, the rate of change of $y(t)$ (and consequently the numerical solution x_n) will be less than that of the exact solution $x(t)$ of the original problem—the numerical solution will then have a tendency to underestimate the magnitude of the true solution under these circumstances. The opposite conclusion will hold when $df/dy < 0$.

To obtain more detailed information we suppose that f is a linear function:

$$f(y) = \lambda y.$$

The modified equation becomes

$$y'(t) = \mu y(t), \qquad \mu = \lambda(1 - \tfrac{1}{2}\lambda h). \quad (13.6)$$

Defining the GE $\hat{e}_n = y(t_n) - x_n$ we find, using (13.3) and $x_{n+1} = (1 + \lambda h)x_n$, that

$$\hat{e}_{n+1} = (1 + \lambda h)\hat{e}_n + \widehat{T}_{n+1}, \quad (13.7)$$

with $\hat{e}_0 = 0$. This is essentially the same as the recurrence (2.15) for the GE for Euler's method, so the proof of Theorem 2.4 can be adapted to prove second-order convergence. That is, $\hat{e}_n = \mathcal{O}(h^2)$ (see Exercise 13.1).

Thus, (13.6) is a suitable modified equation—when h is sufficiently small, x_n is closer to $y(t_n)$ than it is to $x(t_n)$, since

$$\hat{e}_n = y(t_n) - x_n = \mathcal{O}(h^2), \quad \text{while} \quad e_n = x(t_n) - x_n = \mathcal{O}(h).$$

This is borne out in Figure 13.1, where we show the solution of the original problem $x(t)$ (solid curve), the solution of the modified equation $y(t)$ (dashed curve), and the numerical solution x_n with $h = 0.1$ (circles), $\lambda = -5$ (left), and $\lambda = 5$ (right).

We easily calculate in this example that $x(t) = e^{\lambda t}$ and $y(t) = e^{\mu t}$. When $\lambda \in \mathbb{R}$ it can be shown that $\mu < \lambda$ so long as $1 + \frac{1}{2}\lambda h > 0$ (which will always be the case if h is sufficiently small). Hence, if $\lambda < 0$, $y(t)$ (and therefore also x_n) will *decay faster* than $x(t)$ (see Figure 13.1, left). Contrariwise, when $\lambda > 0$, $y(t)$ (and x_n) will *increase more slowly* than $x(t)$ (see Figure 13.1, right)—in both cases one could say that Euler's method introduces too much damping. It is possible, as described in Exercise 13.3, to derive a modified equation that approximates the numerical method to higher orders. The solutions of such a method of order 3 are shown as dotted curves in Figure 13.1.

By retaining the first two terms in the binomial expansion[2] of $(1 + \frac{1}{2}\lambda h)^{-1}$,

$$\frac{\lambda}{1 + \frac{1}{2}\lambda h} = \lambda(1 - \frac{1}{2}\lambda h) + \mathcal{O}(h^2),$$

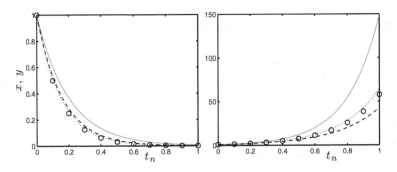

Fig. 13.1 The circles show the numerical solutions for Example 13.1 with $h = 0.1$, $\lambda = -5$ (left), and $\lambda = 5$ (right). The solid curve shows the solution $x(t)$ of the original ODE $x'(t) = \lambda x(t)$, the dashed curve $y(t)$ the solution of the modified Equation (13.6) and the dotted curve the solution of the modified equation of order 3 given in Exercise 13.3

---

[2]We use $p = -1$ in $(1 + z)^p = 1 + pz + \frac{1}{2}p(p - 1)z^2 + \dots$ (which is convergent for $|z| < 1$) and retain only the first two terms. See, for example, [12, 64].

we are led to an alternative modified equation

$$y'(t) = \mu y(t), \qquad \mu = \frac{\lambda}{1 + \frac{1}{2}\lambda h}, \tag{13.8}$$

with which the numerical method is also consistent of second order. This illustrates the non-uniqueness of modified equations. It also allows us to demonstrate the important principal that one *cannot* deduce stability properties of a numerical method by analysing its modified equation(s). Here, when $\lambda < 0$, the solutions to the original modified Equation (13.6) decay to zero for all $h > 0$, since $\mu < 0$. For the alternative modified Equation (13.8), $\mu < 0$ only for those step sizes h for which $1 + \frac{1}{2}h\lambda > 0$. Thus the two possible modified equations have quite different behaviours when h is too large. So the concept is only relevant for sufficiently small h. □

Example 13.2

Use modified equations to compare the behaviour of forward and backward Euler methods for solving the logistic equation $x'(t) = 2x(t)\big(1 - x(t)\big)$ with initial condition $x(0) = 0.1$.

With $f(y) = 2y(1 - y)$, the modified equation (13.5) for Euler's method becomes

$$y'(t) = \big[1 - hy(1 - 2y)\big]y(1 - y), \tag{13.9}$$

while that for the backward Euler method is (see Exercise 13.5)

$$y'(t) = \big[1 + hy(1 - 2y)\big]y(1 - y), \tag{13.10}$$

the initial condition being $y(0) = 0.1$ in both cases. The right-hand sides of both these equations are positive for $0 < y < 1$ and $h < 1$, so the corresponding IVPs have monotonically increasing solutions.

For $0.5 < y < 1$ the solution $y(t)$ of (13.9) satisfies $y'(t) < y(1 - y)$, so the solution of the modified equation (and, therefore, the solution of the forward Euler method) increases more slowly than the exact solution $x(t)$, while, for $0.5 < y < 1$, $y'(t) > y(1 - y)$ and the numerical solution grows more quickly. These properties are reversed for Equation (13.10) and the backward Euler method. These deductions are confirmed by the numerical results shown in Figure 13.2 with $h = 0.3$ and $h = 0.15$. □

In our next example we stay with Euler's method, but this time it is applied to a system of ODEs. The steps involved in the construction of a modified system of equations are similar to those in Example 13.1, except that vector quantities are involved.

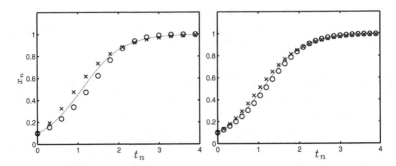

Fig. 13.2 Numerical solutions of the logistic equation with the forward Euler method (circles) and backward Euler method (crosses) of Example 13.2 with $h = 0.3$ (left) and $h = 0.15$ (right). Also shown is the exact solution $x(t)$ of the IVP (solid curve)

Example 13.3

Euler's method applied to the IVP

$$u'(t) = -v(t), \qquad v'(t) = u(t),$$
$$u(0) = 1, \qquad v(0) = 0 \tag{13.11}$$

leads to

$$u_{n+1} = u_n - hv_n, \quad v_{n+1} = v_n + hu_n, \qquad n = 0, 1, \ldots,$$
$$u_0 = 1, \quad v_0 = 0, \tag{13.12}$$

and the numerical solutions with $h = 1/2$ are displayed on the left of Figure 7.3. Derive a modified system of equations that will capture the behaviour of the numerical solution.

We suppose that the modified equation is a system of two ODEs with dependent variables $x(t)$ and $y(t)$. The LTE of the given method is, therefore,

$$\widehat{\boldsymbol{T}}_{n+1} = \begin{bmatrix} x(t+h) - x(t) + hy(t) \\ y(t+h) - y(t) - hx(t) \end{bmatrix}, \quad t = nh, \tag{13.13}$$

which, by Taylor expansion, becomes

$$\widehat{\boldsymbol{T}}_{n+1} = h \begin{bmatrix} x'(t) + \frac{1}{2}hx''(t) + y(t) \\ y'(t) + \frac{1}{2}hy''(t) - x(t) \end{bmatrix} + \mathcal{O}(h^3). \tag{13.14}$$

We now suppose that the modified equations take the form

$$x'(t) = -y(t) + ha(x, y),$$
$$y'(t) = \ \ x(t) + hb(x, y),$$

where the functions $a(x, y)$ and $b(x, y)$ are to be determined. Differentiating these with respect to t gives

$$x''(t) = -y'(t) + \mathcal{O}(h) = -x(t) + \mathcal{O}(h),$$
$$y''(t) = \quad x'(t) + \mathcal{O}(h) = -y(t) + \mathcal{O}(h).$$

Substitution into (13.14) then leads to

$$\widehat{T}_{n+1} = h^2 \begin{bmatrix} a(x, y) - \frac{1}{2}x(t) \\ b(x, y) - \frac{1}{2}y(t) \end{bmatrix} + \mathcal{O}(h^3).$$

Therefore, $\widehat{T}_{n+1} = \mathcal{O}(h^3)$ on choosing $a(x, y) = \frac{1}{2}x$ and $b(x, y) = \frac{1}{2}y$. Our modified system of equations is, therefore,

$$\begin{aligned} x'(t) &= -y(t) + \tfrac{1}{2}hx(t), \\ y'(t) &= \quad x(t) + \tfrac{1}{2}hy(t). \end{aligned} \tag{13.15}$$

They can be written in matrix-vector form as (see also Exercise 13.6)

$$\boldsymbol{x}'(t) = \widehat{A}\boldsymbol{x}(t), \quad \boldsymbol{x}(t) = \begin{bmatrix} x(t) \\ y(t) \end{bmatrix}, \quad \widehat{A} = \begin{bmatrix} \frac{1}{2}h & -1 \\ 1 & \frac{1}{2}h \end{bmatrix}. \tag{13.16}$$

When $x(t)$ and $y(t)$ satisfy these ODEs the LTE (13.13) is of order $\mathcal{O}(h^3)$ and so Euler's method must be convergent to $\boldsymbol{x}(t)$ of order 2. That is, the solutions of (13.12) satisfy

$$u_n = x(t_n) + \mathcal{O}(h^2), \quad v_n = y(t_n) + \mathcal{O}(h^2).$$

In order to use these modified equations to explain the behaviour of numerical solutions observed in Example 7.6 (see Figure 7.3, left), we observe that the eigenvalues of the matrix A in (13.16) are given by

$$\lambda_\pm = \tfrac{1}{2}h \pm i.$$

These have (small) positive real parts, which means that the solutions in the phase plane will spiral outwards. This behaviour can be quantified without having to solve the modified system; it can be deduced (the details are left to Exercise 13.8) that

$$x^2(t) + y^2(t) = e^{ht}. \tag{13.17}$$

The curve described by this equation is a spiral and is shown in Figure 13.3 as a dashed line that accurately predicts the behaviour of the numerical solution (shown as dots) when $h = 1/3$ (such a large step size is used for illustrative purposes). □

The behaviour of Euler's method in the previous example was clearly inappropriate for dealing with an oscillatory problem (characterized by solutions

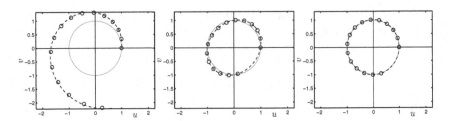

Fig. 13.3 Left: the numerical solutions for Example 13.3 for $0 \le t \le 5$. The dots show the solution of (13.12) with $h = 1/3$ in the u-v phase plane; the solid curve is the circular trajectory of the original IVP and the dashed line the solution of the modified Equation (13.16). In the centre and on the right are shown the corresponding results for Examples 13.4 and 13.5

forming closed curves in the phase plane). We show in the next two examples how small modifications of the method lead to a dramatic improvement in performance.

In the first variation of Euler's method, the usual "forward Euler" (FE) method is applied to the first ODE of the system (13.11) while the backward version (BE) is applied to the second equation.

Example 13.4 (The Symplectic Euler Method)

Derive a modified system of equations that will describe the behaviour of solutions of the method

$$
\begin{aligned}
\text{FE} &: u_{n+1} = u_n - hv_n, \\
\text{BE} &: v_{n+1} = v_n + hu_{n+1},
\end{aligned}
\tag{13.18}
$$

for the IVP (13.11) from the previous example.

Following the same steps as the previous example it can be shown that, instead of (13.15), we arrive at

$$
\left.
\begin{aligned}
x'(t) &= -y(t) + \tfrac{1}{2}hx(t) \\
y'(t) &= x(t) - \tfrac{1}{2}hy(t)
\end{aligned}
\right\},
\tag{13.19}
$$

which differs from (13.15) in that the sign of $y(t)$ on the right-hand side of the second equation has changed. It is now possible to deduce that the trajectories in the phase plane lie on one of the family of ellipses

$$
x^2(t) - hx(t)y(t) + y^2(t) = \text{constant}
\tag{13.20}
$$

(see Exercise 13.9). In this particular case the initial conditions $x(0) = 1$, $y(0) = 0$ fix the constant term to be 1. This ellipse is shown in Figure 13.3

(centre) as a dashed curve. The numerically computed points (u_n, v_n) lie, to within graphical accuracy, exactly on this ellipse. As $h \to 0$ the ellipse collapses to the circle $x^2(t) + y^2(t) = 1$ (shown as a solid curve), which is the trajectory followed by the exact solution of the original equations (see Example 13.1). Because the numerical solutions follow closed orbits the method is well suited to integration of the system over long time intervals. This result would not be significant if it held only for this linear system of differential equations, since it can be solved exactly and there is no practical need for a numerical method. However, the method generalizes quite simply to nonlinear situations, as described in Exercise 13.11. This method is also discussed in a more general context in Section 15.3.   □

The symplectic Euler method (13.18) offers a significant improvement over the standard Euler method but it is still only a first-order accurate method. The second variation of Euler's method leads to a second-order method. In its basic form it begins by applying the symplectic Euler method with a step size $h/2$ and then repeats the process with the order of the ODEs reversed. This removes the bias present in the symplectic Euler method (FE is always applied before BE). The use of half step sizes $h/2$ necessitates the introduction of quantities such as $u_{n-\frac{1}{2}}$ and $u_{n+\frac{1}{2}}$ (known as "half-integer" values) which approximate solutions, respectively, at times $t = (n-\frac{1}{2})h$ and $(n+\frac{1}{2})h$, midway between the points $t = t_{n-1}, t = t_n$ and $t = t_{n+1}$ on the temporal grid.

Example 13.5 (The Störmer–Verlet Method)

Derive a modified system of equations that will describe the behaviour of solutions of the method

$$
\begin{aligned}
\text{FE}: \quad & u_{n+\frac{1}{2}} = u_n - \tfrac{1}{2}hv_n, \\
\text{BE}: \quad & v_{n+\frac{1}{2}} = v_n + \tfrac{1}{2}hu_{n+\frac{1}{2}} \\
\text{FE}: \quad & v_{n+1} = v_{n+\frac{1}{2}} + \tfrac{1}{2}hu_{n+\frac{1}{2}} \\
\text{BE}: \quad & u_{n+1} = u_{n+\frac{1}{2}} - \tfrac{1}{2}hv_{n+1}
\end{aligned}
\left.\begin{aligned} \\ \\ \end{aligned}\right\} v_{n+1} = v_n + hu_{n+\frac{1}{2}}, \qquad (13.21)
$$

for the IVP (13.11) from the previous example.

As indicated, the middle two stages may be combined into one. Further computational savings can be achieved by also combining the last stage of one step with the first step of the next stage. Thus, for computational purposes, the algorithm involves:

1. $u_{\frac{1}{2}} = u_0 - \tfrac{1}{2}hv_0$ and $v_1 = v_0 + hu_{\frac{1}{2}}$.

2. $u_{n+\frac{1}{2}} = u_{n-\frac{1}{2}} + hv_n$ and $v_{n+1} = v_n + hu_{n+\frac{1}{2}}$ for $n = 1, 2, \ldots$.

3. Should the values of u and v be required at time $t = t_{n+1}$:

$$u_{n+1} = u_{n+\frac{1}{2}} - \frac{1}{2}hv_{n+1}.$$

Thus, v is computed at integer nodes and u at half-integer nodes.

For the purposes of analysis, all half-integer values are eliminated from (13.18), so the update from u_n, v_n to u_{n+1}, v_{n+1} may be written as

$$\begin{bmatrix} u_{n+1} \\ v_{n+1} \end{bmatrix} = \begin{bmatrix} u_n \\ v_n \end{bmatrix} + h \begin{bmatrix} -v_n \\ u_n \end{bmatrix} - \frac{1}{2}h^2 \begin{bmatrix} u_n \\ v_n \end{bmatrix} + \frac{1}{4}h^3 \begin{bmatrix} v_n \\ 0 \end{bmatrix}. \tag{13.22}$$

Before embarking on the construction of modified equations we check the LTE of the method. The original ODE system is expressed in matrix-vector form as

$$\boldsymbol{u}'(t) = A\boldsymbol{u}(t), \qquad A = \begin{bmatrix} 0 & -1 \\ 1 & 0 \end{bmatrix}$$

and we observe that $A^2 = -I$, $A^3 = -A$ so that

$$\boldsymbol{u}''(t) = A\boldsymbol{u}'(t) = A^2\boldsymbol{u}(t) = -\boldsymbol{u}(t), \qquad \boldsymbol{u}'''(t) = -\boldsymbol{u}'(t) = -A\boldsymbol{u}(t).$$

The Taylor expansion

$$\boldsymbol{u}(t+h) = \boldsymbol{u}(t) + h\boldsymbol{u}'(t) + \frac{1}{2}h^2\boldsymbol{u}''(t) + \frac{1}{3!}h^3\boldsymbol{u}'''(t) + \mathcal{O}(h^4)$$

therefore becomes

$$\boldsymbol{u}(t+h) = \left(I + hA - \frac{1}{2}h^2I - \frac{1}{3!}h^3A\right)\boldsymbol{u}(t) + \mathcal{O}(h^4).$$

Equation (13.22) can be written in matrix-vector form as

$$\boldsymbol{u}_{n+1} = \left(I + hA - \frac{1}{2}h^2I + \frac{1}{4}h^3 \begin{bmatrix} 0 & 1 \\ 0 & 0 \end{bmatrix}\right)\boldsymbol{u}_n,$$

so, under the localizing assumption $\boldsymbol{u}_n = \boldsymbol{u}(t_n)$, we have

$$\boldsymbol{u}(t_{n+1}) - \boldsymbol{u}_{n+1} = \mathcal{O}(h^3),$$

showing that the Störmer–Verlet method is consistent of order 2.

We next seek a matrix B so that the numerical solution is consistent of order three with the modified system

$$\boldsymbol{x}'(t) = (A + h^2 B)\boldsymbol{x}(t). \tag{13.23}$$

It follows by successive differentiation that

$$\boldsymbol{x}''(t) = (A + h^2 B)\boldsymbol{x}'(t) = (A + h^2 B)^2 \boldsymbol{x}(t) + \mathcal{O}(h^4)$$
$$= A^2 \boldsymbol{x}(t) + \mathcal{O}(h^2) = -\boldsymbol{x}(t) + \mathcal{O}(h^2)$$

and $x'''(t) = -Ax(t) + \mathcal{O}(h^2)$. Using these in the Taylor expansion of $x(t+h)$ we find

$$
\begin{aligned}
x(t+h) &= x(t) + hx'(t) + \tfrac{1}{2}h^2 x''(t) + \tfrac{1}{3!}h^3 x'''(t) + \mathcal{O}(h^4) \\
&= x(t) + h(A + h^2 B)x(t) \\
&\quad - \tfrac{1}{2}h^2 - x(t) - \tfrac{1}{3!}h^3 Ax(t) + \mathcal{O}(h^4) \\
&= \left[I + hA - \tfrac{1}{2}h^2 I - h^3(\tfrac{1}{3!}A + B) \right] x(t) + \mathcal{O}(h^4).
\end{aligned}
$$

Now, under the localizing assumption that $u_n = x(t_n)$, we find

$$
x(t_{n+1}) - u_{n+1} = h^3 \left(-\tfrac{1}{3!}A + B - \tfrac{1}{4} \begin{bmatrix} 0 & 1 \\ 0 & 0 \end{bmatrix} \right) x(t_n) + \mathcal{O}(h^4).
$$

The right-hand side will be of fourth order if

$$
B = \tfrac{1}{3!}A + \tfrac{1}{4} \begin{bmatrix} 0 & 1 \\ 0 & 0 \end{bmatrix} = \tfrac{1}{12} \begin{bmatrix} 0 & -1 \\ 2 & 0 \end{bmatrix}.
$$

The Störmer–Verlet method is, therefore, consistent of order three with the modified system (13.23). It then follows (see Exercise 13.10) that the components of $x(t)$ lie on the ellipse

$$
(1 + \tfrac{1}{6}h^2)x^2(t) + (1 + \tfrac{1}{12}h^2)y^2(t) = \text{constant} \tag{13.24}
$$

and, from the initial conditions, the constant $= 1 + \tfrac{1}{6}h^2$. This is shown as a dashed curve in Figure 13.3 (right) when $h = 1/3$ and is virtually indistinguishable from the circle on which the exact solution $u(t)$ lies.

It was shown in Example 7.6 that the solutions of the trapezoidal rule lie on precisely the same circle as the exact solution of the IVP. This would appear to give it an advantage over the symplectic Euler method (13.18) and the Störmer–Verlet method (13.21). However, counting against the trapezoidal rule are the facts that (a) it is implicit and, therefore computationally expensive in a nonlinear context, and (b) the exact "conservation of energy" property does not generalize to nonlinear problems as it does for the other methods.   □

13.3 A Two-Step Method

The examples thus far have sought to find a modified equation with which the method has a higher order of consistency than with the original ODE. Our final example has a different nature and shows that a method applied to a scalar problem may also be a consistent approximation of a system.

Example 13.6 (The Mid-Point Rule)

Derive a modified system of equations that can be used to explain the behaviour of the mid-point rule (see Example 6.12) when it is used to solve the IVP $u'(t) = u(t) - u^2(t)$, $u(0) = 0.1$. The numerical solution with $h = 0.1$ is shown in Figure 13.4 (left) and is seen to have periodic bursts of activity while the exact solution is effectively constant for $t > 5$.

The mid-point rule applied to the given ODE leads to the nonlinear difference equation

$$u_{n+2} - u_n = 2hu_{n+1}(1 - u_{n+1}) \tag{13.25}$$

with $u_0 = 0.1$ and we assume that the second initial condition is obtained via Euler's method:

$$u_1 = u_0 + hu_0(1 - u_0).$$

We observe in Figure 13.4 (left) that the numerical solution (dots) is indistinguishable from the exact solution (solid curve) up until about $t = 5$, after which consecutive values of u_n oscillate around the steady state $u = 1$. This suggests that we treat the even- and odd-numbered values of u_n differently. We shall suppose that, $u_n \approx x(t_n)$ for even values of n and that $u_n \approx y(t_n)$ for odd values of n, where the smooth functions $x(t)$ and $y(t)$ will be the solutions of a (yet to be determined) modified system of ODEs.

The LTE will be different depending on the parity of n:

$$T_{n+2} = \begin{cases} x(t_{n+2}) - x(t_n) - 2hy(t_{n+1})\big(1 - y(t_{n+1})\big), & \text{when } n \text{ is even,} \\ y(t_{n+2}) - y(t_n) - 2hx(t_{n+1})\big(1 - x(t_{n+1})\big), & \text{when } n \text{ is odd,} \end{cases} \tag{13.26}$$

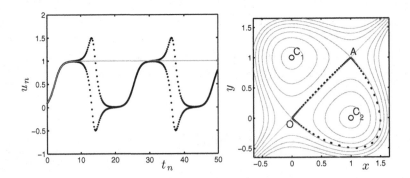

Fig. 13.4 Left: numerical solutions for Example 13.6 with $h = 0.1$ (dots) and the exact solution $u(t)$ (solid curve). Right: the numerical solution (dots) is seen to lie on one of the family (13.28) of ellipses in phase space

and Taylor expansion[3] leads to

$$T_{n+2} = \begin{cases} 2h\big(x'(t_{n+1}) - y(t_{n+1})\big(1 - y(t_{n+1})\big)\big) + \mathcal{O}(h^3), & \text{when } n \text{ is even,} \\ 2h\big(y'(t_{n+1}) - x(t_{n+1})\big(1 - x(t_{n+1})\big)\big) + \mathcal{O}(h^3), & \text{when } n \text{ is odd.} \end{cases}$$

This suggests that the method, which is a one-step map from (u_n, u_{n+1}) to (u_{n+1}, u_{n+2}), is consistent of order 2 with the system

$$\left. \begin{array}{l} x'(t) = y(t) - y^2(t) \\ y'(t) = x(t) - x^2(t) \end{array} \right\}. \tag{13.27}$$

In contrast to the previous examples this system does not have h-dependent coefficients. Since

$$\frac{dy}{dx} = \frac{y'(t)}{x'(t)},$$

we deduce that y, regarded as a function of x, satisfies the separable differential equation

$$(y(t) - y^2(t))\frac{dy}{dx} = x(t) - x^2(t).$$

This can be integrated to give (following factorization)

$$(x - y)\big[\tfrac{1}{2}x + \tfrac{1}{2}y - \tfrac{1}{3}(x^2 + xy + y^2)\big] = \text{constant.} \tag{13.28}$$

These curves are shown in Figure 13.4 (right) with values $-0.15, -0.1, \ldots, 0.15$ for the constant. Superimposed are the points (u_{2m}, u_{2m-1}) showing even- versus odd-numbered values of the solution sequence. The dots leading directly from the origin O to the point marked A$(1, 1)$ correspond to the smooth part of the trajectory up to about $t = 5$. On OA we have $x = y$, which implies that $x = y = u$, and so even- and odd-numbered points both approximate the original solution $u(t)$.

When the solution approaches the steady state we can substitute $u(t) = 1 - \varepsilon v(t)$ into $u'(t) = u(t) - u^2(t)$ to give $v'(t) = -v(t) + \varepsilon v^2(t)$. When ε is small, $u(t)$ is close to 1 and the linearized equation[4] $v'(t) = -v(t)$ indicates that the solution is attracted to $v = 0$, that is $u = 1$, exponentially in time. However, it was shown in Example 6.12 that the mid-point rule cannot be absolutely stable for any $h > 0$. It is this loss of (weak) stability that causes the oscillatory behaviour—corresponding to a trajectory moving in a closed semi-elliptic path in the phase plane. The long-term behaviour of the system (13.27) is the subject of Exercise 12.1.

---

[3]It is more efficient here to Taylor expand each of the quantities $x(t_{n+2})$, $x(t_n)$, $y(t_{n+2})$, $y(t_n)$ about $t = t_{n+1}$ because this avoids having to expand the nonlinear terms.

[4]See Section 12.1 on Long Term Dynamics.

How can this behaviour be reconciled with the fact that, because the mid-point rule is zero-stable and a consistent method of order 2 to the original ODE, its solutions must converge to $u(t)$, the solution of $u'(t) = u(t) - u^2(t)$, $u(0) = 0.1$? In the phase plane it takes an infinite time for the exact solution of the ODEs to reach the point A and convergence of a numerical method is only guaranteed for finite time intervals $[0, t_f]$. Moreover, as $h \to 0$, the time at which the instability sets in tends to infinity and so the motion on any interval $[0, t_f]$ is ultimately free of oscillations.   □

13.4 Postscript

In cases where the modified equations are differential equations with h-dependent coefficients they should tend towards the original differential equations as $h \to 0$. This provides a basic check on derived modified equations. See the article by Griffiths and Sanz-Serna [24] for further examples of modified equations for both ordinary and partial differential equations in a relatively simple setting.

Modified equations are related to the idea of "backward error analysis" in linear algebra that was developed around 1950 (see N.J. Higham [36, Section 1.5]). The motivating idea is that instead of regarding our computed values as an *approximate* solution to the *given* problem, we may regard them as an *exact* solution to a *nearby* problem. We may then try to quantify the concept of "nearby" and study whether the nearby problem inherits properties of the given problem. In our context, the nearby problem arises by adding small terms to the right-hand side of the ODE. An alternative would be to allow the initial conditions to be perturbed, which leads to the concept of *shadowing*. This has been extensively studied in the context of dynamical systems (see, for example, Chow and Vleck [8]).

EXERCISES

13.1.* For Example 13.1, amend the proof of Theorem 2.4 to prove that $\hat{e}_n = \mathcal{O}(h^2)$ (Hint: it is only necessary to take account of the fact that $|\widehat{T}_j| \leq Ch^3$.)

13.2.** Show that the LTE (13.3) is of order $\mathcal{O}(h^3)$ when $y(t)$ is the solution of the alternative modified Equation (13.8). Hence, conclude that x_n is a second-order approximation to $y(t_n)$. Show that the arguments based on (13.6) about the overdamping effects of Euler's method could also be deduced from (13.8).

13.3.** Show that the LTE (13.3) is of order $\mathcal{O}(h^3)$ when $y(t)$ is the solution of the modified equation $y'(t) = \mu y(t)$, where[5]

$$\mu = \lambda\big(1 - \tfrac{1}{2}\lambda h + \tfrac{1}{3}\lambda^2 h^2\big).$$

Hence, conclude that x_n is a third-order approximation to $y(t_n)$.

13.4.** Consider the backward Euler method $(1 - \lambda h)x_{n+1} = x_n$ applied to the ODE $x'(t) = \lambda x(t)$. Show that the LTE is given by

$$\widehat{T}_{n+1} = y(t_n + h) - (1 - \lambda h)^{-1}y(t_n)$$

and is of order $\mathcal{O}(h^3)$ when $y(t)$ satisfies the second-order ODE

$$y'(t) = \lambda(1 + \tfrac{1}{2}\lambda h)y(t).$$

Deduce that $|y(t)| > |x(t)|$ for $t > 0$ when $0 < 1 + \tfrac{1}{2}\lambda h < 1$ and $y(0) = x(0)$.

13.5.** Consider the backward Euler method $x_{n+1} = x_n + hf_{n+1}$ applied to the ODE $x'(t) = f(x(t))$. By writing $x_{n+1} = x_n + \delta_n$, show that

$$f(x_{n+1}) = f(x_n) + h\frac{\mathrm{d}f}{\mathrm{d}x}(x_n)\delta_n + \mathcal{O}(h^2).$$

Use this, together with the expansion (13.4) and the localizing assumption, to deduce the modified equation

$$y'(t) = \left(1 + \tfrac{1}{2}h\frac{\mathrm{d}f}{\mathrm{d}y}(y(t))\right)f(y(t)). \tag{13.29}$$

13.6.** Show that $\boldsymbol{y}'(t) = A(I - \tfrac{1}{2}hA)\boldsymbol{y}(t)$ is a modified equation for Euler's method applied to the linear system of ODEs $\boldsymbol{x}'(t) = A\boldsymbol{x}(t)$, where $\boldsymbol{x}, \boldsymbol{y} \in \mathbb{R}^m$ and A is an $m \times m$ matrix.

Identify an appropriate matrix A for the ODEs in Example 13.3 and verify that the "modified matrix" in (13.16) is $\widehat{A} = A(I - \tfrac{1}{2}hA)$.

---

[5]An alternative way of finding suitable values of μ is to substitute $y(t) = e^{\mu t}$ into (13.3) to give $\widehat{T}_{n+1} = e^{\mu t_n}\big(e^{\mu h} - 1 - \lambda h\big)$. Hence, $\widehat{T}_{n+1} = \mathcal{O}(h^{p+1})$ if μ is chosen so that $e^{\mu h} = 1 + \lambda h + \mathcal{O}(h^{p+1})$. Thus, μ can be obtained by truncating the Maclaurin series expansion of

$$\mu = \frac{1}{h}\log(1 + \lambda h) = \lambda - \tfrac{1}{2}\lambda^2 h + \tfrac{1}{3}\lambda^3 h^2 + \cdots + (-1)^p\frac{1}{p+1}\lambda^{p+1}h^p + \cdots.$$

13.7.*** Use the modified system derived in the previous question to discuss the behaviour of Euler's method applied to $x'(t) = Ax(t)$, when $x(t_0)$ is an eigenvector of A, in the cases where the matrix A is (a) positive definite, (b) negative definite, and (c) skew-symmetric.

13.8.* Use the identity

$$\frac{\mathrm{d}}{\mathrm{d}t}\left(x^2(t) + y^2(t)\right) = 2x(t)x'(t) + 2y(t)y'(t)$$

together with the ODEs (13.15) to prove that

$$\frac{\mathrm{d}}{\mathrm{d}t}\left(x^2(t) + y^2(t)\right) = h\left(x^2(t) + y^2(t)\right),$$

which is a first-order constant-coefficient linear differential equation in the dependent variable $w(t) = x^2(t) + y^2(t)$. Hence, prove that $x(t)$ and $y(t)$ satisfy Equation (13.17).

13.9.** Complete the details leading up to (13.19).

Show that

$$\frac{\mathrm{d}}{\mathrm{d}t}\left(x^2(t) - hx(t)y(t) + y^2(t)\right) = 0$$

and, hence, deduce that Equation (13.20) holds.

13.10.* Deduce that the solutions of the modified system derived in Example 13.5 lie on the family of ellipses (13.24).

13.11.*** Show that the nonlinear oscillator $u''(t) + f(u) = 0$ (cf. Exercise 7.6) is equivalent to the first-order system

$$u'(t) = -v(t), \qquad v'(t) = f(u(t)). \tag{13.30}$$

Suppose that $F(u)$ is such that $f(u) = \dfrac{\mathrm{d}F(u)}{\mathrm{d}u}$. Show that the solutions of this system satisfy

$$\frac{\mathrm{d}}{\mathrm{d}t}\left(2F(u(t)) + v^2(t)\right) = 0$$

and, therefore, lie on the family of curves $v^2(t) + 2F(u(t)) = $ constant. The system (13.30) may be solved numerically by a generalization of the method in Example 13.4:

$$u_{n+1} = u_n - hv_n, \qquad v_{n+1} = v_n + hf(u_{n+1}), \qquad n = 0, 1, \ldots,$$

with $u_0 = 1$ and $v_0 = 0$. Derive the modified system

$$\left. \begin{array}{l} x'(t) = -y(t) + \frac{1}{2}hf(x(t)) \\ y'(t) = \quad f(x(t)) - \frac{1}{2}h\dfrac{\mathrm{d}f}{\mathrm{d}x}(x(t))y(t) \end{array} \right\}.$$

Deduce that

$$\frac{\mathrm{d}}{\mathrm{d}t}\left(2F(x(t)) - hf(x(t))y(t) + y^2(t)\right) = 0$$

and hence that solutions lie on the family of curves

$$2F(x(t)) - hf(x(t))y(t) + y^2(t) = \text{constant}$$

in the x-y phase plane. Verify that these results coincide with those of Example 13.4 when $f(u) = u$.

13.12.*** It follows from Exercise 11.3 that the TS(1) method of Example 11.2 applied to the IVP $x'(t) = -\lambda x(t)$, $x(0) = 1$, leads to

$$x_{n+1} = (1 - h_n\lambda)x_n, \qquad t_{n+1} = t_n + h_n,$$

$$h_n = \left|\frac{2\,\text{tol}}{\lambda^2 x_n}\right|^{1/2}, \qquad t_0 = 0, \qquad x_0 = 1, \tag{13.31}$$

on the assumption that no steps are rejected. Suppose that (t_n, x_n) denotes an approximation to a point on the parameterized curve $(t(s), x(s))$ at $s = s_n$, where $s_n = n\Delta s$ and $\Delta s = \sqrt{2\text{tol}}$ is a constant step size in s. Assuming that $\lambda > 0$, show that the equations (13.31) are a consistent approximation of the IVP

$$\frac{\mathrm{d}}{\mathrm{d}s}x(s) = -\lambda g(s)x(s), \qquad \frac{\mathrm{d}}{\mathrm{d}s}t(s) = g(s),$$

with $x(0) = 1$, $t(0) = 0$, and $g(s) = 1/(\lambda\sqrt{x(s)})$.

Solve these ODEs for $x(s)$ and $t(s)$ and verify that the expected solution is obtained when the parameter s is eliminated. Show that $x(s) = 0$ for $s = 2$ regardless of the value of λ. Deduce that the numerical solution x_n is expected to reach the fixed point $x = 0$ in approximately $\sqrt{2/\text{tol}}$ time steps. Sample numerical results are presented in Figure 13.5 ($\sqrt{2/\text{tol}} = \sqrt{200} \approx 14.14$ and $x_{14} = 0.0046$.

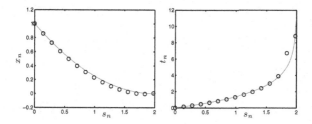

Fig. 13.5 Numerical solution for Exercise 13.12 with $\lambda = 1$ and tol $= 0.01$. Left: (s_n, x_n) (Circles) and $(s, x(s))$ (solid), Right: (s_n, t_n) (circles) and $(s, t(s))$ (solid)

14
Geometric Integration Part I—Invariants

14.1 Introduction

We judge a numerical method by its ability to "approximate" the ODE. It is perfectly natural to

- fix an initial condition,

- fix a time t_f

and ask how closely the method can match $x(t_f)$, perhaps in the limit $h \to 0$. This led us, in earlier chapters, to the concepts of *global error* and *order of convergence*. However, there are other senses in which approximation quality may be studied. We have seen that *absolute stability* deals with long-time behaviour on linear ODEs, and we have also looked at simple long-time dynamics on nonlinear problems with *fixed points*. In this chapter and the next we look at another well-defined sense in which the ability of a numerical method to reproduce the behaviour of an ODE can be quantified—we consider ODEs with a *conservative* nature—that is, certain algebraic quantities remain constant (are conserved) along trajectories. This gives us a taste of a very active research area that has become known as *geometric integration*, a term that, to the best of our knowledge, was coined by Sanz-Serna in his review article [60]. The material in these two chapters borrows heavily from Hairer et al. [26] and Sanz-Serna and Calvo [61].

Throughout both chapters, we focus on autonomous ODEs, where the right-hand side does not depend explicitly upon t, so we have $x'(t) = f(x)$.

D.F. Griffiths, D.J. Higham, *Numerical Methods for Ordinary Differential Equations*,
Springer Undergraduate Mathematics Series, DOI 10.1007/978-0-85729-148-6_14,
© Springer-Verlag London Limited 2010

14.2 Linear Invariants

A simple chemical reaction known as a *reversible isometry* may be written

$$X_1 \underset{k_2 \, X_2}{\overset{k_1 \, X_1}{\rightleftharpoons}} X_2. \tag{14.1}$$

Here, a molecule of chemical species X_1 may spontaneously convert into a molecule of chemical species X_2, and vice versa. The constants k_1 and k_2 reflect the rates at which these two reactions occur. If these two species are not involved in any other reactions, then clearly, the sum of the number of X_1 and X_2 molecules remains constant.

The mass action ODE for this chemical system has the form

$$\left. \begin{aligned} u'(t) &= -k_1 u(t) + k_2 v(t) \\ v'(t) &= k_1 u(t) - k_2 v(t), \end{aligned} \right\} \tag{14.2}$$

where $u(t)$ and $v(t)$ represent the concentrations of X_1 and X_2 respectively. For initial conditions $u(0) = A$ and $v(0) = B$, this ODE has solution

$$\begin{aligned} u(t) &= \frac{k_2(A+B)}{k_1+k_2} + \left[A - \frac{k_2(A+B)}{k_1+k_2} \right] e^{-(k_1+k_2)t}, \\ v(t) &= \frac{k_1(A+B)}{k_1+k_2} + \left[B - \frac{k_1(A+B)}{k_1+k_2} \right] e^{-(k_1+k_2)t}; \end{aligned} \tag{14.3}$$

see Exercise 14.1. Adding gives $u(t) + v(t) = A + B$, showing that the ODE system automatically preserves the sum of the concentrations of species X_1 and X_2.

Applying Euler's method to (14.2) gives us

$$\begin{aligned} u_{n+1} &= u_n - hk_1 u_n + hk_2 v_n, \\ v_{n+1} &= v_n + hk_1 u_n - hk_2 v_n. \end{aligned}$$

Summing these two equations, we find that $u_{n+1} + v_{n+1} = u_n + v_n$. This shows that Euler's method preserves the total concentration, matching the property of the ODE. This is not a coincidence—by developing a little theory, we will show that a more general result holds.

Instead of solving the system (14.2) and combining the solutions, we could simply add the two ODEs, noting that the right-hand sides cancel, to give $u'(t) + v'(t) = 0$; that is,

$$\frac{\mathrm{d}}{\mathrm{d}t} \left(u(t) + v(t) \right) = 0.$$

We may deduce immediately that $u(t) + v(t)$ remains constant along any solution of the ODE. Given a general ODE system $\boldsymbol{x}'(t) = \boldsymbol{f}(\boldsymbol{x})$ with $\boldsymbol{x}(t) \in \mathbb{R}^m$

and $\boldsymbol{f} : \mathbb{R}^m \to \mathbb{R}^m$, we say that there is a *linear invariant* if some linear combination of the solution components is always preserved:

$$\frac{\mathrm{d}}{\mathrm{d}t}\left(c_1 x_1(t) + c_2 x_2(t) + \cdots + c_m x_m(t)\right) = 0,$$

where $c_1, c_2, \ldots, c_m$ are constants. We may write this more compactly as

$$\frac{\mathrm{d}}{\mathrm{d}t}\boldsymbol{c}^{\mathrm{T}}\boldsymbol{x}(t) = 0,$$

where $\boldsymbol{c} \in \mathbb{R}^m$. Because each x_i' is given by $f_i(\boldsymbol{x})$, this is equivalent to

$$c_1 f_1(\boldsymbol{x}) + c_2 f_2(\boldsymbol{x}) + \cdots + c_m f_m(\boldsymbol{x}) = 0,$$

which we may write as

$$\boldsymbol{c}^{\mathrm{T}}\boldsymbol{f}(\boldsymbol{x}) = 0, \qquad \text{for any } \boldsymbol{x} \in \mathbb{R}^m. \tag{14.4}$$

The condition (14.4) asks for $\boldsymbol{f}$ to satisfy a specific geometric property: wherever it is evaluated, the result must always be orthogonal to the fixed vector $\boldsymbol{c}$.

For Euler's method, $\boldsymbol{x}_{n+1} = \boldsymbol{x}_n + h\boldsymbol{f}(\boldsymbol{x}_n)$ and, under the condition (14.4), we have

$$\boldsymbol{c}^{\mathrm{T}}\boldsymbol{x}_{n+1} = \boldsymbol{c}^{\mathrm{T}}\boldsymbol{x}_n + h\boldsymbol{c}^{\mathrm{T}}\boldsymbol{f}(\boldsymbol{x}_n) = \boldsymbol{c}^{\mathrm{T}}\boldsymbol{x}_n, \tag{14.5}$$

so the corresponding linear combination of the components in the numerical solution is also preserved. More generally, since they update from step to step by adding multiples of $\boldsymbol{f}$ values, it is easy to see that any consistent linear multistep method and any Runge–Kutta method will preserve linear invariants of an ODE; see Exercise 14.2.

Linear invariants arise naturally in several application areas.

- In chemical kinetics systems like (14.2), it is common for some linear combination of the individual population counts to remain fixed. (This might not be as simple as maintaining the overall sum; for example, one molecule of species X_1 may combine with one molecule of species X_2 to produce a single molecule of species X_3.)

- In mechanical models, the overall mass of a system may be conserved.

- In stochastic models where an ODE takes the form of a "master equation" describing the evolution of discrete probabilities, the total probability must sum to one.

- In epidemic models where individuals move between susceptible, infected and recovered states, the overall population size may be fixed.

In all cases, we have shown that standard numerical methods automatically inherit the same property. Now it is time to move on to nonlinear invariants.

14.3 Quadratic Invariants

The angular momentum of a free rigid body with centre of mass at the origin can be described by the ODE system [26, Example 1.7, Chapter IV]

$$\left.\begin{array}{l} x_1'(t) = a_1\, x_2(t)\, x_3(t) \\[4pt] x_2'(t) = a_2\, x_3(t)\, x_1(t) \\[4pt] x_3'(t) = a_3\, x_1(t)\, x_2(t) \end{array}\right\}. \tag{14.6}$$

Here, a_1, a_2, and a_3 are constants that depend on the fixed principal moments of inertia, I_1, I_2, and I_3, according to

$$a_1 = \frac{I_2 - I_3}{I_2\, I_3}, \quad a_2 = \frac{I_3 - I_1}{I_3\, I_1}, \quad a_3 = \frac{I_1 - I_2}{I_1\, I_2}.$$

Note that, by construction,

$$a_1 + a_2 + a_3 = 0. \tag{14.7}$$

The quantity $x_1^2(t) + x_2^2(t) + x_3^2(t)$ is an *invariant* for this system; it remains constant over time along any solution. We can check this by direct differentiation, using the form of the ODEs (14.6) and the property (14.7):

$$\begin{aligned} \frac{\mathrm{d}}{\mathrm{d}t}\left(x_1^2 + x_2^2 + x_3^2\right) &= 2\,x_1\, x_1' + 2\,x_2\, x_2' + 2\,x_3\, x_3' \\[4pt] &= 2\,x_1\, a_1\, x_2\, x_3 + 2\,x_2\, a_2\, x_3\, x_1 + 2\,x_3\, a_3\, x_1\, x_2 \\[4pt] &= 2\,x_1\, x_2\, x_3\,(a_1 + a_2 + a_3) \\[4pt] &= 0. \end{aligned} \tag{14.8}$$

So, every solution of (14.6) lives on a sphere; that is,

$$x_1^2(0) + x_2^2(0) + x_3^2(0) = R^2$$

implies that

$$x_1^2(t) + x_2^2(t) + x_3^2(t) = R^2$$

for all $t > 0$. Exercise 14.5 asks you to check that the quantity

$$\frac{x_1^2}{I_1} + \frac{x_2^2}{I_2} + \frac{x_3^2}{I_3} \tag{14.9}$$

is also an invariant for this system. It follows that solutions are further constrained to live on fixed ellipsoids. So, overall, any solution lies on a curve formed by the intersection of a sphere and an ellipsoid. Examples of such curves are shown in Figure 14.1, which we will return to soon.

Generally, we will say that an ODE system $x'(t) = f(x)$ has a *quadratic invariant* if the function

$$x^T C x \tag{14.10}$$

is preserved, where $C \in \mathbb{R}^{m \times m}$ is a symmetric matrix. For the rigid body example (14.6) we have just seen that there are two quadratic invariants, with

$$C = \begin{bmatrix} 1 & 0 & 0 \\ 0 & 1 & 0 \\ 0 & 0 & 1 \end{bmatrix} \quad \text{and} \quad C = \begin{bmatrix} 1/I_1 & 0 & 0 \\ 0 & 1/I_2 & 0 \\ 0 & 0 & 1/I_3 \end{bmatrix}. \tag{14.11}$$

By definition, if (14.10) is an invariant then its time derivative is zero, giving

$$0 = \frac{d}{dt} \left(x^T C x \right) = x'^T C x + x^T C x' = 2 x^T C x' = 2 x^T C f(x),$$

where we have used $C = C^T$ and $x'(t) = f(x)$. This shows that $x^T C x$ is a quadratic invariant if and only if the function f satisfies

$$x^T C f(x) = 0, \qquad \text{for any } x \in \mathbb{R}^m. \tag{14.12}$$

The *implicit mid-point rule* is defined by

$$x_{n+1} = x_n + h f \left(\frac{x_n + x_{n+1}}{2} \right). \tag{14.13}$$

This method achieves the commendable feat of preserving quadratic invariants of the ODE for any choice of stepsize h. To see this, we will simplify the notation by letting

$$f_n^{\mathrm{mid}} := f \left(\frac{x_n + x_{n+1}}{2} \right).$$

Then one step of the implicit midpoint rule produces an approximation x_{n+1} for which

$$x_{n+1}^T C x_{n+1} = \left(x_n + h f_n^{\mathrm{mid}} \right)^T C \left(x_n + h f_n^{\mathrm{mid}} \right)$$
$$= x_n^T C x_n + 2 h x_n^T C f_n^{\mathrm{mid}} + h^2 f_n^{\mathrm{mid}\,T} C f_n^{\mathrm{mid}}.$$

From (14.13), we may write

$$x_n = \frac{x_n + x_{n+1}}{2} - \tfrac{1}{2} h f_n^{\mathrm{mid}},$$

and so

$$x_{n+1}^T C x_{n+1} = x_n^T C x_n + 2 h \left(\frac{x_n + x_{n+1}}{2} - \tfrac{1}{2} h f_n^{\mathrm{mid}} \right)^T C f_{\mathrm{mid}} + h^2 f_n^{\mathrm{mid}\,T} C f_n^{\mathrm{mid}}$$

$$= x_n^T C x_n + 2 h \left(\frac{x_n + x_{n+1}}{2} \right)^T C f_n^{\mathrm{mid}}.$$

But the last term on the right disappears when we invoke the condition (14.12). Hence,

$$x_{n+1}^T C x_{n+1} = x_n^T C x_n,$$

confirming that the numerical method shares the quadratic invariant.

By contrast, applying Euler's method

$$x_{n+1} = x_n + h f_n, \tag{14.14}$$

we find, with f_n denoting $f(x_n)$, that

$$\begin{aligned} x_{n+1}^T C x_{n+1} &= \left(x_n + h f_n^T \right) C \left(x_n + h f_n \right) \\ &= x_n^T C x_n + 2h x_n^T C f_n + h^2 f_n^T C f_n \\ &= x_n^T C x_n + h^2 f_n^T C f_n. \end{aligned}$$

The extra term $h^2 f_n^T C f_n$ will be nonzero in general, and hence Euler's method does not preserve quadratic invariants of the ODE. In fact, for the two types of matrix C in (14.11)[1] we have $f_n^T C f_n > 0$ for any $f_n \neq 0$. So, on the rigid body problem (14.6) Euler's method produces an approximation that drifts outwards onto increasingly large spheres and ellipsoids. This is a more general case of the behaviour observed in Examples 7.6 and 13.3.

These results are confirmed experimentally in Figure 14.1, which is inspired by Hairer et al. [26, Figure 1.1, Chapter IV, page 96]. Here, we illustrate the behaviour of the implicit mid-point rule (left) and Euler's method (right) on the rigid body problem (14.6). In each picture, the dots denote the numerical approximations, plotted in the x_1, x_2, x_3 phase space. In this case we have $x_1(0)^2 + x_2(0)^2 + x_3(0)^2 = 1$, and the corresponding unit spheres are indicated in the pictures. The solid curves mark the intersection between the sphere and various ellipsoids of the form (14.9) with $I_1 = 2$, $I_2 = 1$ and $I_3 = 2/3$. Any exact solution to the ODE remains on such an intersection, and we see that the numerical solution provided by the implicit mid-point rule shares this property. The Euler approximation, however, shown on the right, is seen to violate these constraints by spiralling away from the surface of the sphere.

Example 14.1 (Kepler Problem)

The Kepler two-body problem describes the motion of two planets. With appropriate normalization, this ODE system takes the form [26, 61]

$$\begin{aligned} p_1'(t) &= -\frac{q_1}{\left(q_1^2 + q_2^2 \right)^{3/2}}, & q_1'(t) &= p_1, \\ p_2'(t) &= -\frac{q_2}{\left(q_1^2 + q_2^2 \right)^{3/2}}, & q_2'(t) &= p_2, \end{aligned} \tag{14.15}$$

---

[1] And, more generally, for any *symmetric positive definite* matrix C.

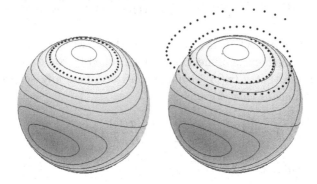

Fig. 14.1 Dots show numerical approximations for the implicit mid-point rule (left) and Euler's method (right) on the rigid body problem (14.6). Intersections between the unit sphere and ellipsoids of the form (14.9) are marked as solid lines. (This figure is based on Figure 1.1 in Chapter IV of the book by Hairer et al. [26])

where $q_1(t)$ and $q_2(t)$ give the position of one planet moving in a plane relative to the other at time t.

This system has a quadratic invariant of the form

$$q_1 p_2 - q_2 p_1, \tag{14.16}$$

which corresponds to angular momentum. This can be checked by direct differentiation:

$$\frac{\mathrm{d}}{\mathrm{d}t}(q_1 p_2 - q_2 p_1) = q_1' p_2 + q_1 p_2' - q_2' p_1 - q_2 p_1'$$

$$= p_1 p_2 - \frac{q_1 q_2}{(q_1^2 + q_2^2)^{3/2}} - p_2 p_1 + \frac{q_1 q_2}{(q_1^2 + q_2^2)^{3/2}} = 0.$$

However, the system also has a nonlinear invariant given by the *Hamiltonian* function

$$H(p_1, p_2, q_1, q_2) = \tfrac{1}{2}\left(p_1^2 + p_2^2\right) - \frac{1}{\sqrt{q_1^2 + q_2^2}}, \tag{14.17}$$

(see Exercise 14.11), so called because the original ODEs (14.15) may be written in the form

$$p_1'(t) = -\frac{\partial}{\partial q_1} H(p_1, p_2, q_1, q_2), \qquad q_1'(t) = \frac{\partial}{\partial p_1} H(p_1, p_2, q_1, q_2),$$

$$p_2'(t) = -\frac{\partial}{\partial q_2} H(p_1, p_2, q_1, q_2), \qquad q_2'(t) = \frac{\partial}{\partial p_2} H(p_1, p_2, q_1, q_2). \tag{14.18}$$

Hamiltonian ODEs are considered in the next chapter[2]. □

---

[2]We will consider problems where $H : \mathbb{R}^2 \to \mathbb{R}$, but the same ideas extend to the more general case of $H : \mathbb{R}^{2d} \to \mathbb{R}$.

14.4 Modified Equations and Invariants

Suppose now that the ODE system $\boldsymbol{x}'(t) = \boldsymbol{f}(\boldsymbol{x})$, with $\boldsymbol{x}(t) \in \mathbb{R}^m$ and $\boldsymbol{f} : \mathbb{R}^m \to \mathbb{R}^m$, has a general nonlinear invariant defined by some smooth function $\mathcal{F} : \mathbb{R}^m \to \mathbb{R}$, so that $\mathcal{F}(\boldsymbol{x}(t))$ remains constant along any solution. By the chain rule, this means that, for any $\boldsymbol{x}(t)$,

$$0 = \frac{\mathrm{d}}{\mathrm{d}t} \mathcal{F}(\boldsymbol{x}(t)) = \sum_{i=1}^{m} \frac{\partial \mathcal{F}(\boldsymbol{x}(t))}{\partial x_i} \left(\boldsymbol{f}(\boldsymbol{x}(t)) \right)_i .$$

We may write this as

$$(\nabla \mathcal{F}(\boldsymbol{x}))^{\mathrm{T}} \boldsymbol{f}(\boldsymbol{x}) = 0, \qquad \text{for any } \boldsymbol{x} \in \mathbb{R}^m, \tag{14.19}$$

where $\nabla \mathcal{F}$ denotes the vector of partial derivatives of $\mathcal{F}$. This is a direct generalization of the linear invariant case (14.4) and the quadratic invariant case (14.12).

Suppose we have a numerical method of order p that is able to preserve the same invariant. So, from any initial value $\boldsymbol{x}(0)$, if we take N steps to reach time $t_{\mathrm{f}} = Nh$ then the global error satisfies

$$\boldsymbol{x}_N - \boldsymbol{x}(t_{\mathrm{f}}) = \mathcal{O}(h^p), \tag{14.20}$$

and, because the invariant is preserved, we have

$$\mathcal{F}(\boldsymbol{x}_N) = \mathcal{F}(\boldsymbol{x}(0)). \tag{14.21}$$

Now, following the ideas in Chapter 13, we know that it is generally possible to construct a modified IVP of the form

$$\boldsymbol{y}'(t) = \boldsymbol{f}(\boldsymbol{y}) + h^p \boldsymbol{g}(\boldsymbol{y}), \qquad \boldsymbol{y}(0) = \boldsymbol{x}(0), \tag{14.22}$$

such that the numerical method approximates the solution $\boldsymbol{y}(t)$ even more accurately than it approximates the original solution $\boldsymbol{x}(t)$. In general, by adding the extra term on the right-hand side in (14.22), we are able to increase the order by one, so that

$$\boldsymbol{x}_N - \boldsymbol{y}(t_{\mathrm{f}}) = \mathcal{O}(h^{p+1}). \tag{14.23}$$

We will show now that this modified equation *automatically inherits the invariance property of the ODE and numerical method*. This can be done through a contradiction argument. Suppose that the modified equation (14.22) does not satisfy the invariance property (14.19). Then there must be at least one point $\boldsymbol{x}^\star \in \mathbb{R}^m$ where $(\nabla \mathcal{F}(\boldsymbol{x}^\star))^{\mathrm{T}} \boldsymbol{g}(\boldsymbol{x}^\star) \neq 0$. By switching from $\mathcal{F}$ to $-\mathcal{F}$ if necessary, we may assume that this nonzero value is positive and then, by continuity, there must be a region B containing $\boldsymbol{x}^\star$ for which

$$(\nabla \mathcal{F}(\boldsymbol{x}))^{\mathrm{T}} \boldsymbol{g}(\boldsymbol{x}) > C, \qquad \text{for all } \boldsymbol{x} \in B,$$

where $C > 0$ is a constant. If we choose x^* as our initial condition and let $[0, t_f]$ be a time interval over which $y(t)$ remains inside the region B, then,

$$\frac{\mathrm{d}}{\mathrm{d}t}\mathcal{F}(y(t)) = h^p \left(\nabla\mathcal{F}(y(t))\right)^{\mathrm{T}} g(x) > Ch^p, \qquad \text{for } t \in [0, t_f].$$

After integrating both sides with respect to t from 0 to t_f, we find

$$|\mathcal{F}(y(t_f)) - \mathcal{F}(y(0))| > Ct_f h^p. \tag{14.24}$$

On the other hand, since the method preserves the invariant, we have $\mathcal{F}(x_N) = \mathcal{F}(y(0))$, so

$$\mathcal{F}(y(t_f)) - \mathcal{F}(y(0)) = \mathcal{F}(y(t_f)) - \mathcal{F}(x_N).$$

Because $\mathcal{F}$ is smooth and we are working in a compact set B, the difference on the right may be bounded by a multiple of the difference in the arguments[3]. So, using (14.23), we have

$$\mathcal{F}(y(t_f)) - \mathcal{F}(y(0)) = \mathcal{O}(h^{p+1}).$$

However, this contradicts the bound (14.24). So we conclude that the modified equation must preserve the invariant.

14.5 Discussion

It is natural to ask which standard numerical methods can preserve quadratic invariants. In the case of Runge–Kutta methods, (9.5)) and (9.6), there is a very neat result. Cooper [10] proved that the condition

$$b_i a_{ij} + b_j a_{ji} = b_i b_j, \qquad \text{for all } i, j = 1, \dots, s, \tag{14.25}$$

characterizes successful methods. Of course, it is also possible to construct *ad hoc* methods that deal with specific classes of ODEs with quadratic, or more general, invariants.

The result that a modified equation inherits invariants that are shared by the ODE and the numerical method can be established very generally for one-step methods, and the same principle applies for other properties, including *reversibility, volume preservation, fixed point preservation* and, as we study in the next chapter, *symplecticness*. The proof by contradiction argument that we used was taken from Gonzalez et al. [22], where those other properties are also considered.

---

[3]More precisely, we may assume that a Lipschitz condition $|\mathcal{F}(x) - \mathcal{F}(y)| \leq L\|x - y\|$ holds for $x, y \in B$.

The modified equation concept plays a central role in the modern study of geometric integration. By optimizing over the number of extra terms in the modified equation, it is possible to show that the difference between a numerical solution and its closest modified equation can be bounded by $C \exp(-D/h)$ over a time interval $0 \le t \le E/h$, where C, D and E are constants. This is an *exponentially small* bound that applies over *arbitrarily large time*; a very rare phenomenon in numerical ODEs![4] Establishing the existence of a modified equation with the same structure as the original ODE is often a key step in proving further positive results, such as mild growth of the global error as a function of time.

Finally, we should add that not all properties of an ODE and numerical method are automatically inherited by a modified equation; see Exercise 14.12 for an illustration.

EXERCISES

14.1.** Confirm that (14.3) satisfies the ODE system (14.2). Also check that this solution gives $u(t) + v(t) \equiv A + B$. What happens to this solution as $t \to \infty$? Explain the result intuitively in terms of the reaction rate constants k_1 and k_2.

14.2.** By generalizing the analysis for the Euler case in (14.5), show that any consistent linear multistep method and any Runge–Kutta method will preserve linear invariants of an ODE. You may assume that the starting values for a k-step LMM satisfy $c^{\mathrm{T}} \eta_j = K$, $j = 0 : k - 1$, for some constant K when c is any vector such that equation (14.4) holds.

14.3.** Show that the second-order Taylor series method, TS(2), from Chapter 3, applied to (14.2) takes the form

$$\begin{bmatrix} u_{n+1} \\ v_{n+1} \end{bmatrix} = \begin{bmatrix} u_n \\ v_n \end{bmatrix} + h\left(1 - \tfrac{1}{2}h(k_1 + k_2)\right) \begin{bmatrix} -k_1 & k_2 \\ k_1 & -k_2 \end{bmatrix} \begin{bmatrix} u_n \\ v_n \end{bmatrix}.$$

Confirm that this method also preserves the linear invariant.

14.4.** For the system (14.2) show that $u'(t) = Au(t)$, where A may be written as the outer product

$$A = \begin{bmatrix} -1 \\ 1 \end{bmatrix} \begin{bmatrix} k_1, -k_2 \end{bmatrix}.$$

---

[4]We should note, however, that the bound does not involve the solution of the original ODE. Over a long time interval the modified equation might not remain close to the underlying problem.

Deduce that $A^j = (-k_1 - k_2)^{j-1} A$ $(j = 1, 2, 3, \dots)$ and hence generalise the previous exercise to the $TS(p)$ method for $p > 2$.

14.5.* By differentiating directly, as in (14.8), show that the quadratic expression (14.9) is an invariant for the ODE system (14.6).

14.6.* For (14.6) confirm that (14.12) holds for the two choices of matrix C in (14.11).

14.7.** Show that the implicit mid-point rule (14.13) may be regarded as an implicit Runge–Kutta method with Butcher tableau (as defined in Section 9.2) given by

$$
\begin{array}{c|c}
\frac{1}{2} & \frac{1}{2} \\
\hline
 & 1
\end{array} \; .
$$

14.8.*** By following the analysis for the implicit mid-point rule, show that the implicit Runge–Kutta method with Butcher tableau

$$
\begin{array}{c|cc}
\frac{1}{2} - \frac{\sqrt{3}}{6} & \frac{1}{4} & \frac{1}{4} - \frac{\sqrt{3}}{6} \\
\frac{1}{2} + \frac{\sqrt{3}}{6} & \frac{1}{4} + \frac{\sqrt{3}}{6} & \frac{1}{4} \\
\hline
 & \frac{1}{2} & \frac{1}{2}
\end{array}
$$

called the *order 4 Gauss method*, also preserves quadratic invariants of an ODE system for any step size h.

14.9.* The backward Euler method produces $x_{n+1} = x_n + h f(x_{n+1})$, instead of (14.14). In this case show that

$$
x_n^T C x_n = x_{n+1}^T C x_{n+1} + h^2 f_n^T C f_n
$$

when (14.12) holds for the ODE. Hence, explain how the picture on the right in Figure 14.1 would change if Euler's method were replaced by backward Euler.

14.10.* Show that the invariant (14.16) may be written in the form

$$
\begin{bmatrix} p_1 & p_2 & q_1 & q_2 \end{bmatrix} C \begin{bmatrix} p_1 \\ p_2 \\ q_1 \\ q_2 \end{bmatrix} ,
$$

where $C \in \mathbb{R}^{4 \times 4}$ is symmetric, so that it fits into the format required in (14.10).

14.11.* By directly differentiating, show that H in (14.17) is an invariant for the Kepler problem (14.15).

14.12.*** (Based on Gonzalez et al. [22, Section 4.6].) Consider the scalar ODE $x'(t) = f(x(t))$, with $f(x) = -x^3$. Letting $\mathcal{F}(x) = x^4/4$, use the chain rule to show that

$$\frac{d}{dt}\mathcal{F}(x(t)) = -\left(f(x(t))\right)^2.\tag{14.26}$$

Deduce that

$$\lim_{t\to\infty} x(t) = 0, \qquad \text{for any } x(0).$$

For this ODE, rather than being an invariant, $\mathcal{F}$ is a *Lyapunov function* that always decreases along each non-constant trajectory. For the backward Euler method $x_{n+1} = x_n - hx_{n+1}^3$, by using the Taylor series with remainder it is possible to prove a discrete analogue of (14.26),

$$\frac{\mathcal{F}(x_{n+1}) - \mathcal{F}(x_n)}{h} \leq -\left(\frac{x_{n+1} - x_n}{h}\right)^2.$$

It follows that, given any h,

$$\lim_{n\to\infty} x_n = 0, \qquad \text{for any } x(0).$$

Show that

$$y'(t) = -y^3(t) + \frac{3h}{2}y^5(t)$$

is a modified equation for backward Euler on this ODE, and also deduce that

$$\lim_{t\to\infty} |y(t)| = \infty, \qquad \text{when } y(0) > \sqrt{\frac{2}{3h}}.$$

This gives an example of a qualitative property that is shared by an ODE and a numerical method for all h, but is not inherited by a corresponding modified equation.

15

Geometric Integration Part II—Hamiltonian Dynamics

This chapter continues our study of geometric features of ODEs. We look at Hamiltonian problems, which possess the important property of symplecticness. As in the previous chapter our emphasis is on

- showing that methods must be carefully chosen if they are to possess the correct geometric property, and

- using the idea of modified equations to explain the qualitative behaviour of numerical methods.

15.1 Symplectic Maps

As a lead-in to the topics of Hamiltonian ODEs and symplectic maps, we begin with some key geometric concepts. Figure 15.1 depicts a parallelogram with vertices at $(0,0)$, (a,b), $(a+c, b+d)$, and (c,d). The area of the parallelogram can be written as $|ad - bc|$; see Exercise 15.1. If we remove the absolute value sign, then the remaining expression $ad - bc$ is positive if the vertices are listed in clockwise order, otherwise it is negative. Hence, characterizing the parallelogram in terms of the two vectors

$$u = \begin{bmatrix} a \\ b \end{bmatrix} \quad \text{and} \quad v = \begin{bmatrix} c \\ d \end{bmatrix},$$

D.F. Griffiths, D.J. Higham, *Numerical Methods for Ordinary Differential Equations*,
Springer Undergraduate Mathematics Series, DOI 10.1007/978-0-85729-148-6_15,
© Springer-Verlag London Limited 2010

we may define the *oriented area*, $\text{area}_o(\boldsymbol{u}, \boldsymbol{v})$, to be $ad - bc$. Equivalently, we may write

$$\text{area}_o(\boldsymbol{u}, \boldsymbol{v}) = \boldsymbol{u}^{\mathrm{T}} J \boldsymbol{v}, \tag{15.1}$$

where $J \in \mathbb{R}^{2 \times 2}$ has the form

$$J = \begin{bmatrix} 0 & 1 \\ -1 & 0 \end{bmatrix}.$$

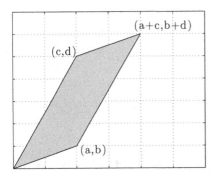

Fig. 15.1 Parallelogram

Now, given a matrix $A \in \mathbb{R}^{2 \times 2}$, we may ask whether the oriented-area is preserved under the linear mapping $\boldsymbol{u} \mapsto A\boldsymbol{u}$ and $\boldsymbol{v} \mapsto A\boldsymbol{v}$. From (15.1), we will have $\text{area}_o(\boldsymbol{u}, \boldsymbol{v}) = \text{area}_o(A\boldsymbol{u}, A\boldsymbol{v})$ if and only if $\boldsymbol{u}^{\mathrm{T}} A^{\mathrm{T}} J A \boldsymbol{v} = \boldsymbol{u}^{\mathrm{T}} A \boldsymbol{v}$. Hence, the linear mapping guarantees to preserve oriented area if, and only if,

$$A^{\mathrm{T}} J A = J. \tag{15.2}$$

Figure 15.2 illustrates this idea. Here, the parallelogram on the right is found by applying a linear mapping to $\boldsymbol{u}$ and $\boldsymbol{v}$, and, since we have chosen a matrix A for which $A^{\mathrm{T}} J A = J$, the oriented area is preserved. We also note that the condition (15.2) is equivalent to $\det(A) = 1$; see Exercise 15.5. However, we prefer to use the formulation (15.2) as it is convenient algebraically and, for our purposes, it extends more naturally to the case of higher dimensions.

Adopting a more general viewpoint, we may consider any smooth nonlinear mapping $\boldsymbol{g} : \mathbb{R}^2 \to \mathbb{R}^2$. When is $\boldsymbol{g}$ *oriented area preserving*? In other words, if we take a two-dimensional region and apply the map $\boldsymbol{g}$ to every point, this will give us a new region in $\mathbb{R}^2$. Under what conditions will the two regions always have the same oriented area? Fixing on a point $\boldsymbol{x} \in \mathbb{R}^2$, we imagine a small parallelogram formed by the vectors $\boldsymbol{x} + \boldsymbol{\varepsilon}$ and $\boldsymbol{x} + \boldsymbol{\delta}$, where $\boldsymbol{\varepsilon}, \boldsymbol{\delta} \in \mathbb{R}^2$ are arbitrary but small, as indicated in the left-hand picture of Figure 15.3. Placing the origin at $\boldsymbol{x}$, we know from (15.1) that the oriented area of this parallelogram is given by

$$\boldsymbol{\varepsilon}^{\mathrm{T}} J \boldsymbol{\delta}. \tag{15.3}$$

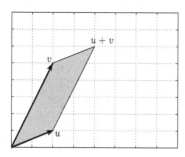
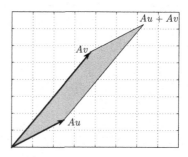

Fig. 15.2 The parallelogram on the right was formed by using a matrix A to map the vectors $\boldsymbol{u}$ and $\boldsymbol{v}$ that define the edges of the parallelogram on the left. In this case $\det(A) = 1$, so the area is preserved

If we apply the map $\boldsymbol{g}$ to this parallelogram then we get a region in $\mathbb{R}^2$, as indicated in the right-hand picture. Because $\boldsymbol{\varepsilon}$ and $\boldsymbol{\delta}$ are small, the sides of this region can be approximated by straight lines, and the new region can be approximated by a parallelogram. Also, by linearizing the map, as explained in Appendix C, we may approximate the locations of the vertices $\boldsymbol{g}(\boldsymbol{x} + \boldsymbol{\varepsilon})$ and $\boldsymbol{g}(\boldsymbol{x} + \boldsymbol{\delta})$ using (C.3) to obtain

$$\boldsymbol{g}(\boldsymbol{x} + \boldsymbol{\varepsilon}) \approx \boldsymbol{g}(\boldsymbol{x}) + \frac{\partial \boldsymbol{g}}{\partial \boldsymbol{x}}(\boldsymbol{x})\,\boldsymbol{\varepsilon},$$

$$\boldsymbol{g}(\boldsymbol{x} + \boldsymbol{\delta}) \approx \boldsymbol{g}(\boldsymbol{x}) + \frac{\partial \boldsymbol{g}}{\partial \boldsymbol{x}}(\boldsymbol{x})\,\boldsymbol{\delta}.$$

Here the Jacobian $\partial \boldsymbol{g}/\partial \boldsymbol{x}$ is the matrix of partial derivatives, see (C.4). Placing the origin at $\boldsymbol{g}(\boldsymbol{x})$, the oriented area of this region may be approximated by the oriented area of the parallelogram, to give

$$\boldsymbol{\varepsilon}^{\mathrm{T}} \left(\frac{\partial \boldsymbol{g}}{\partial \boldsymbol{x}}\right)^{\mathrm{T}} J \left(\frac{\partial \boldsymbol{g}}{\partial \boldsymbol{x}}\right) \boldsymbol{\delta}. \tag{15.4}$$

Equating the oriented areas (15.3) and (15.4) gives the relation

$$\boldsymbol{\varepsilon}^{\mathrm{T}} J \boldsymbol{\delta} = \boldsymbol{\varepsilon}^{\mathrm{T}} \left(\frac{\partial \boldsymbol{g}}{\partial \boldsymbol{x}}\right)^{\mathrm{T}} J \left(\frac{\partial \boldsymbol{g}}{\partial \boldsymbol{x}}\right) \boldsymbol{\delta}.$$

Now the directions $\boldsymbol{\varepsilon}$ and $\boldsymbol{\delta}$ are arbitrary, and we would like this area preservation to hold for any choices. This leads us to the condition

$$J = \left(\frac{\partial \boldsymbol{g}}{\partial \boldsymbol{x}}\right)^{\mathrm{T}} J \left(\frac{\partial \boldsymbol{g}}{\partial \boldsymbol{x}}\right). \tag{15.5}$$

The final step is to argue that any region in $\mathbb{R}^2$ can be approximated to any desired level of accuracy by a collection of small parallelograms. Hence, this

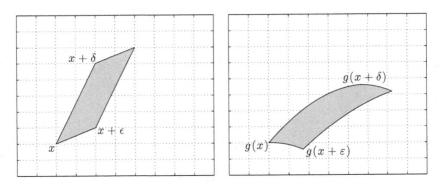

Fig. 15.3 The region on the right illustrates the effect of applying a smooth nonlinear map g to the set of points making up the parallelogram on the left. If the original parallelogram is small, then the map is well approximated by its linearisation (equivalently, the new region is well approximated by a parallelogram).

condition is enough to guarantee area preservation in general. These arguments can be made rigorous and a smooth map g satisfying condition (15.5) does indeed preserve oriented area.

We will use the word *symplectic* to denote maps that satisfy (15.5).[1]

15.2 Hamiltonian ODEs

With a unit mass and unit gravitational constant the motion of a pendulum can be modelled by the ODE system

$$\begin{aligned} p'(t) &= -\sin q(t), \\ q'(t) &= p(t). \end{aligned} \qquad (15.6)$$

Here, the *position coordinate*, $q(t)$, denotes the angle between the rod and the vertical at time t.

Figure 15.4 shows solutions of this system in the q, p phase space. In each case, the initial condition is marked with a circle. We have also indicated the

---

[1] For maps in $\mathbb{R}^{2d}$ where $d > 1$, the concept of symplecticness is stronger than simple preservation of volume—it is equivalent to preserving the sum of the oriented areas of projections on to two-dimensional subspaces. However, the characterization (15.5) remains valid when the matrix J is generalized appropriately to higher dimension, and the algebraic manipulations that we will use can be carried through automatically.

vector field for this system—an arrow shows the direction and size of the tangent vector for a solution passing through that point. In other words, an arrow at the point (q, p) has direction given by the vector $(p, -\sin q)$ and length proportional to $\sqrt{p^2 + \sin^2 q}$. There is a fixed point at $p = q = 0$, corresponding to the pendulum resting in its lowest position, and an unstable fixed point at $q = \pi$, $p = 0$, corresponding to the pendulum resting precariously in its highest position. For any solution, the function

$$H(p, q) = \tfrac{1}{2}p^2 - \cos q \tag{15.7}$$

satisfies

$$\frac{\mathrm{d}}{\mathrm{d}t} H(p, q) = pp' + (\sin q)q' = -p \sin q + (\sin q)p = 0.$$

Hence, $H(p, q)$ remains constant along each solution. In fact, for $-1 < H < 1$, we have the typical "grandfather clock" *libration*—solutions where the angle q oscillates between two fixed values $\pm c$. One such solution is shown towards the centre of Figure 15.4. For $H > 1$, the angle q varies monotonically; here, the pendulum continually swings beyond the vertical. Two of these solutions are shown in the upper and lower regions of the figure. The intermediate case, $H = 1$, corresponds to the unstable resting points and the *separatrices* that connect them are marked with dashed lines (see Exercise 15.8). We also note that the problem is 2π periodic in the angle q.

The pendulum system (15.6) may be written in terms of the function $H : \mathbb{R}^2 \to \mathbb{R}^2$ in (15.7) as

$$\begin{aligned}
\frac{\mathrm{d}p}{\mathrm{d}t} &= -\frac{\partial H}{\partial q}, \\
\frac{\mathrm{d}q}{\mathrm{d}t} &= \frac{\partial H}{\partial p}.
\end{aligned} \tag{15.8}$$

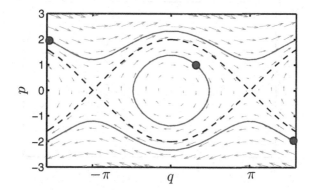

Fig. 15.4 Vector field and particular solutions of the pendulum system (15.6).

It turns out that many other important ODE systems can also be written in this special form (or higher-dimensional analogues, as in the Kepler example (14.18)). Hence, this class of problems is worthy of special attention. Generally, any system of the form (15.8) is called a *Hamiltonian* ODE and $H(p, q)$ is referred to as the corresponding *Hamiltonian function*. The invariance of H that we saw for the pendulum example carries through, because, by the chain rule,

$$\frac{\mathrm{d}}{\mathrm{d}t} H(p, q) = \frac{\partial H}{\partial p} p' + \frac{\partial H}{\partial q} q' = -\frac{\partial H}{\partial p} \frac{\partial H}{\partial q} + \frac{\partial H}{\partial q} \frac{\partial H}{\partial p} = 0.$$

However, in addition to preserving the Hamiltonian function, these problems also have a symplectic character. To see this, we need to introduce the *time t flow map*, ψ_t, which takes the initial condition $p(0) = p_0$, $q(0) = q_0$ and maps it to the solution $p(t), q(t)$; that is,

$$\psi_t \left(\begin{bmatrix} p_0 \\ q_0 \end{bmatrix} \right) = \begin{bmatrix} p(t) \\ q(t) \end{bmatrix}.$$

Once we have fixed t, this defines a map $\psi_t : \mathbb{R}^2 \to \mathbb{R}^2$, and we may therefore investigate whether this map preserves oriented area.

Before carrying this out, the results of a numerical experiment are displayed in Figure 15.5. The effect of applying ψ_t to all the points in the shaded circular disc with centre at $q = 0$, $p = 1.6$ and radius 0.55 is shown for $t = 1$ and $t = 2$ and $t = 3$. In other words, taking every point in the disk as an initial value for the ODE system, we show where the solution lies at times $t = 1, 2, 3$. At time $t = 1$, the disk is rotated clockwise and squashed into a distorted egg shape. At time $t = 2$ the region is stretched into a teardrop by the vector field, and by $t = 3$ it has taken on a tadpole shape.

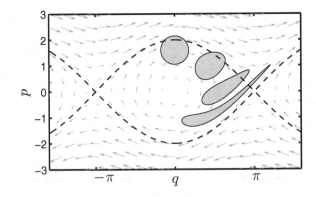

Fig. 15.5 Illustration of the area-preserving property for the flow of the pendulum system (15.6)

It would be reasonable to conjecture from Figure 15.5 that the time t flow map ψ_t is symplectic—the shaded regions appear to have equal areas. To provide a proof we use the approach used in the previous section with $g(x) \equiv \psi_t(x)$ and $x = [p, q]^T$. We first note that, for $t = 0$, ψ_t collapses to the identity map, so its Jacobian (C.4) is the identity matrix, which trivially satisfies (15.5). Next, we consider how the right-hand side of (15.5) evolves over time. The Jacobian matrix (C.7) for (15.6) has the form

$$\begin{bmatrix} 0 & -\cos q \\ 1 & 0 \end{bmatrix}.$$

Then, letting

$$P(t) = \frac{\partial \psi_t}{\partial x_0}$$

denote the matrix of partial derivatives, the variational equation for this matrix has the form

$$P'(t) = \begin{bmatrix} 0 & -\cos q \\ 1 & 0 \end{bmatrix} P(t)$$

(see (C.6)). It follows that

$$\frac{d}{dt}\left(P(t)^T J P(t)\right) = P'(t)^T J P(t) + P(t)^T J P'(t)$$

$$= P(t)^T \begin{bmatrix} 0 & 1 \\ -\cos q & 0 \end{bmatrix} \begin{bmatrix} 0 & 1 \\ -1 & 0 \end{bmatrix} P(t)$$

$$+ P(t)^T \begin{bmatrix} 0 & 1 \\ -1 & 0 \end{bmatrix} \begin{bmatrix} 0 & -\cos q \\ 1 & 0 \end{bmatrix} P(t)$$

$$= P(t)^T \left(\begin{bmatrix} -1 & 0 \\ 0 & -\cos q \end{bmatrix} + \begin{bmatrix} 1 & 0 \\ 0 & \cos q \end{bmatrix} \right) P(t)$$

$$= 0. \tag{15.9}$$

Having first established that (15.5) holds at time $t = 0$, we have now shown that the right-hand side of (15.5) does not change over time; hence, the condition holds for any time t. That is, the time t flow map, ψ_t, for the pendulum equation is symplectic for any t.

The general case, where $H(p, q)$ is any Hamiltonian function in (15.8), can be handled without much more effort. The ODE then has a Jacobian of the form

$$\begin{bmatrix} -H_{pq} & -H_{qq} \\ H_{pp} & H_{qp} \end{bmatrix}.$$

Here, we have adopted the compact notation where, for example, H_{pq} denotes $\partial^2 H / \partial p \partial q$. This Jacobian may be written as $J^{-1} \nabla^2 H$, where

$$J^{-1} = \begin{bmatrix} 0 & -1 \\ 1 & 0 \end{bmatrix} \quad \text{and} \quad \nabla^2 H = \begin{bmatrix} H_{pp} & H_{qp} \\ H_{pq} & H_{qq} \end{bmatrix}.$$

Here, the symmetric matrix $\nabla^2 H$ is called the *Hessian* of H. The computations leading up to (15.9) then generalise to

$$\frac{\mathrm{d}}{\mathrm{d}t}\left(P(t)^{\mathrm{T}} J P(t)\right) = P'(t)^{\mathrm{T}} J P(t) + P(t)^{\mathrm{T}} J P'(t)$$
$$= P(t)^{\mathrm{T}} \nabla^2 H J^{-\mathrm{T}} J P(t) + P(t)^{\mathrm{T}} \nabla^2 H P(t) = 0,$$

since $J^{-\mathrm{T}} J = -I$. So the argument used for the pendulum ODE extends to show that any Hamiltonian ODE has a symplectic time t flow map, ψ_t.

15.3 Approximating Hamiltonian ODEs

We will describe two examples of approximating systems of Hamiltonian ODEs.

Example 15.1

Show that map generated when Euler's method is applied the pendulum ODEs (15.6) is not symplectic.

Euler's method corresponds to the map

$$\left[\begin{array}{c} p_{n+1} \\ q_{n+1} \end{array}\right] = \left[\begin{array}{c} p_n - h \sin q_n \\ q_n + h p_n \end{array}\right]. \tag{15.10}$$

We find that

$$\left[\begin{array}{cc} \dfrac{\partial p_{n+1}}{\partial p_n} & \dfrac{\partial p_{n+1}}{\partial q_n} \\ \dfrac{\partial q_{n+1}}{\partial p_n} & \dfrac{\partial q_{n+1}}{\partial q_n} \end{array}\right] = \left[\begin{array}{cc} 1 & -h \cos q_n \\ h & 1 \end{array}\right],$$

and the right-hand side of (15.5) is

$$\left[\begin{array}{cc} 0 & 1 + h^2 \cos q_n \\ -1 - h^2 \cos q_n & 0 \end{array}\right].$$

Since this is not the identity matrix, the map arising from Euler's method on this ODE is not symplectic for any step size $h > 0$.

Figure 15.6 (top) shows what happens when the exact flow map in Figure 15.5 is replaced by the Euler map (15.10) with step size $h = 1$. The disk of initial values clearly starts to expand its area, before collapsing into a thin smear. Figure 15.6 (bottom) repeats the experiment, this time using the backward Euler method. Here, the region spirals towards the origin, and the area appears to shrink. Exercise 15.7 asks you to check that the corresponding map is not symplectic. □

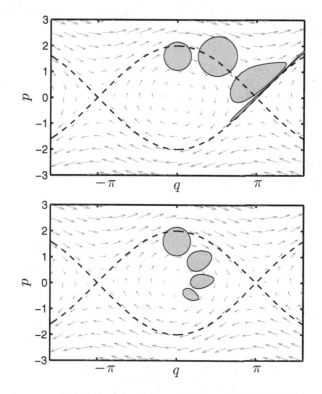

Fig. 15.6 Area evolution for the forward Euler method (top) and backward Euler method (bottom) applied to the pendulum system (15.6)

Example 15.2

Extend the symplectic Euler method from Example 13.4 to the general Hamiltonian system of ODEs (15.8) and verify that the resulting map is symplectic.

The *symplectic Euler* method can be regarded as a combination of forward and backward Euler methods where the updates are implicit in the p variable and explicit in the q variable. When applied to the system (15.8) it leads to

$$p_{n+1} = p_n - h\frac{\partial H}{\partial q}(p_{n+1}, q_n),$$

$$q_{n+1} = q_n + h\frac{\partial H}{\partial p}(p_{n+1}, q_n). \tag{15.11}$$

As its name suggests, the method is symplectic. To confirm this property we first compute partial derivatives to obtain

$$\frac{\partial p_{n+1}}{\partial p_n} = 1 - hH_{pq}(p_{n+1}, q_n)\frac{\partial p_{n+1}}{\partial p_n},$$

$$\frac{\partial p_{n+1}}{\partial q_n} = -hH_{qq}(p_{n+1}, q_n) - hH_{pq}(p_{n+1}, q_n)\frac{\partial p_{n+1}}{\partial q_n},$$

$$\frac{\partial q_{n+1}}{\partial p_n} = hH_{pp}(p_{n+1}, q_n)\frac{\partial p_{n+1}}{\partial p_n},$$

$$\frac{\partial q_{n+1}}{\partial q_n} = 1 + hH_{qp}(p_{n+1}, q_n) + hH_{pp}(p_{n+1}, q_n)\frac{\partial p_{n+1}}{\partial q_n}.$$

Collecting these together, we have

$$\begin{bmatrix} 1 + hH_{pq}(p_{n+1}, q_n) & 0 \\ -hH_{pp}(p_{n+1}, q_n) & 1 \end{bmatrix} \begin{bmatrix} \dfrac{\partial p_{n+1}}{\partial p_n} & \dfrac{\partial p_{n+1}}{\partial q_n} \\ \dfrac{\partial q_{n+1}}{\partial p_n} & \dfrac{\partial q_{n+1}}{\partial q_n} \end{bmatrix} = \begin{bmatrix} 1 & -hH_{qq}(p_{n+1}, q_n) \\ 0 & 1 + hH_{qp}(p_{n+1}, q_n) \end{bmatrix}.$$

We may solve to obtain the Jacobian explicitly, and verify that (15.5) holds. Alternatively, taking determinants directly in this expression leads to the equivalent condition

$$\det\left(\begin{bmatrix} \dfrac{\partial p_{n+1}}{\partial p_n} & \dfrac{\partial p_{n+1}}{\partial q_n} \\ \dfrac{\partial q_{n+1}}{\partial p_n} & \dfrac{\partial q_{n+1}}{\partial q_n} \end{bmatrix}\right) = 1.$$

Figure 15.7 shows how the results in Figure 15.6 change when the symplectic Euler method is used. Comparing the regions in Figure 15.7 with those in Figure 15.5 for the exact time t flow map, we see that the symplectic Euler method is not producing highly accurate approximations—this is to be expected with a relatively large stepsize of $h = 1$. However, the method gives a much better qualitative feel for the behaviour of this ODE than the non-symplectic explicit or implicit Euler methods. In Section 15.4 we will briefly look at the question of how to quantify the benefits of symplecticness.   □

An important practical point is that the symplectic Euler method (15.11) is explicit on the pendulum ODE (15.6). Given p_n and q_n, we may compute $p_{n+1} = p_n - h\sin q_n$ and then $q_{n+1} = q_n + hp_{n+1}$. Hence this method, which is first-order accurate, is just as cheap to compute with as the basic Euler method. This explicit nature carries through more generally when the Hamiltonian function $H(p, q)$ has the *separable* form

$$H(p, q) = T(p) + V(q), \tag{15.12}$$

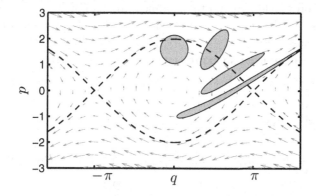

Fig. 15.7 Area evolution for the symplectic Euler method applied to the pendulum ODE (15.6)

which often arises in mechanics, with T and V representing the *kinetic* and *potential* energies, respectively. In this case, symplectic Euler takes the form

$$p_{n+1} = p_n - hV'(q_n),$$
$$q_{n+1} = q_n + hT'(p_{n+1}). \qquad (15.13)$$

15.4 Modified Equations

We will now analyse the map (15.13) arising when the symplectic Euler method (15.11) is applied to the separable Hamiltonian problem

$$p'(t) = -V'(q),$$
$$q'(t) = T'(p), \qquad (15.14)$$

corresponding to (15.12), in the spirit of Chapter 13 and Section 14.4. Special cases of these equations were studied in examples given in these earlier chapters. We look for a modified equation of the form

$$u'(t) = -V'(v) + hA(u, v),$$
$$v'(t) = T'(u) + hB(u, v). \qquad (15.15)$$

We would like one step of the symplectic Euler method applied to the original problem to match this new problem even more closely. In other words, we would like the map (15.13) to match the time h flow map of the modified problem (15.15) more closely than it matches the original problem (15.14).

We therefore Taylor expand the modified problem over one step and choose the functions A and B to match the Taylor expansion of the numerical method as closely as possible.

To expand u, we need an expression for the second derivative. We have

$$
\begin{aligned}
\frac{d^2 u}{dt^2} &= \frac{d}{dt} u' \\
&= \frac{d}{dt} \left(-V'(v) + hA(u,v) \right) \\
&= \frac{\partial}{\partial u} \left(-V'(v) + hA(u,v) \right) \frac{d}{dt} u + \frac{\partial}{\partial v} \left(-V'(v) + hA(u,v) \right) \frac{d}{dt} v \\
&= \mathcal{O}(h) - \left(V''(v) + \mathcal{O}(h) \right) \left(T'(u) + \mathcal{O}(h) \right) \\
&= -V''(v)T'(u) + \mathcal{O}(h).
\end{aligned}
$$

Hence, a Taylor expansion may be written

$$
\begin{aligned}
u(t_n + h) &= u(t_n) + hu'(t_n) + \tfrac{1}{2}h^2 u''(t_n) + \mathcal{O}(h^3) \\
&= u(t_n) - hV'(v(t_n)) \\
&\quad + h^2 \left(A(u(t_n), v(t_n)) - \tfrac{1}{2}V''(v(t_n))T'(u(t_n)) \right) + \mathcal{O}(h^3).
\end{aligned}
$$

So, to match the map (15.13) up to $\mathcal{O}(h^3)$ we require

$$
A(p,q) = \tfrac{1}{2}V''(q)T'(p). \tag{15.16}
$$

Similarly,

$$
\begin{aligned}
\frac{d^2 v}{dt^2} &= \frac{d}{dt} v' \\
&= \frac{d}{dt} \left(T'(u) + hB(u,v) \right) \\
&= \frac{\partial}{\partial u} \left(T'(u) + hB(u,v) \right) \frac{d}{dt} u + \frac{\partial}{\partial v} \left(T'(u) + hB(u,v) \right) \frac{d}{dt} v \\
&= -T''(u)V'(v) + \mathcal{O}(h).
\end{aligned}
$$

So a Taylor expansion for v is

$$
\begin{aligned}
v(t_n + h) &= v(t_n) + hv'(t_n) + \tfrac{1}{2}h^2 v''(t_n) + \mathcal{O}(h^3) \\
&= v(t_n) + hT'(u(t_n)) \\
&\quad + h^2 \left(B(u(t_n), v(t_n)) - \tfrac{1}{2}T''(u(t_n))V'(v(t_n)) \right) + \mathcal{O}(h^3).
\end{aligned}
$$

$$\tag{15.17}$$

Now, because the second component of the map (15.13) involves p_{n+1}, we must expand to obtain

$$
\begin{aligned}
q_{n+1} &= q_n + hT'(p_{n+1}) \\
&= q_n + hT'(p_n - hV'(q_n)) \\
&= q_n + h\left(T'(p_n) - T''(p_n)hV'(q_n) + \mathcal{O}(h^2)\right).
\end{aligned}
\tag{15.18}
$$

Matching this expansion with (15.17) up to $\mathcal{O}(h^3)$ requires

$$
h^2 B(p,q) - \tfrac{1}{2}h^2 T''(p)V'(q) = -h^2 T''(p)V'(q),
$$

which rearranges to

$$
B(p,q) = -\tfrac{1}{2}T''(p)V'(q).
\tag{15.19}
$$

Inserting our expressions (15.16) and (15.19) into (15.15), we obtain the modified equation

$$
\begin{aligned}
u' &= -V'(v) + \tfrac{1}{2}hV''(v)T'(u), \\
v' &= T'(u) - \tfrac{1}{2}hT''(u)V'(v).
\end{aligned}
\tag{15.20}
$$

It is straightforward to check that this has the form (15.8) of a Hamiltonian problem, with

$$
H(p,q) = T(p) + V(q) - \tfrac{1}{2}hT'(p)V'(q).
\tag{15.21}
$$

So the symplectic Euler method may be accurately approximated by a *modified equation that shares the Hamiltonian structure of the underlying ODE.*

In Figure 15.8 we combine the information from Figure 15.5 concerning the exact flow map (dark shading) and Figure 15.7 concerning the h map of the symplectic Euler method (light shading). Between these two sets of regions, we have inserted in medium shading with a dashed white border the regions arising from the time $t = 1$ flow map of the modified equation (15.20) We see that, even for this relatively large value of h, the modified equation does a visibly better job than the original ODE of describing the numerical approximation.

Applying the Euler method to the separable problem (15.14) produces the map

$$
\begin{aligned}
p_{n+1} &= p_n - hV'(q_n), \\
q_{n+1} &= q_n + hT'(p_n).
\end{aligned}
$$

This differs from the symplectic Euler method (15.13) in that p_n, rather than p_{n+1}, appears on the right-hand side of the second equation. It follows that if we repeat the procedure above and look for a modified problem of the form (15.15), then the same $A(p,q)$ arises, but, because the $-h^2 T''(u)V'(v)$ term is missing from the right-hand side of (15.18), we obtain

$$
B(p,q) = \tfrac{1}{2}T''(p)V'(q).
$$

The resulting modified problem for Euler's method is

$$
\begin{aligned}
u' &= -V'(v) + \tfrac{1}{2}hV''(v)T'(u), \\
v' &= T'(u) + \tfrac{1}{2}hT''(u)V'(v).
\end{aligned}
\tag{15.22}
$$

This is not of the Hamiltonian form (15.8).

In summary, the symplectic Euler method has produced a modified equation of Hamiltonian form, and the regular, nonsymplectic, Euler method has not.

We may move on to the next level of modified equation for the symplectic Euler method, by looking for functions $C(u, v)$ and $D(u, v)$ in

$$
\begin{aligned}
u' &= -V'(v) + \tfrac{1}{2}hV''(v)T'(u) + h^2 C(u, v), \\
v' &= T'(u) - \tfrac{1}{2}hT''(u)V'(v) + h^2 D(u, v),
\end{aligned}
\tag{15.23}
$$

such that the relevant expansions match to $\mathcal{O}(h^4)$. After some effort (see Exercise 15.14) we find that this modified equation is also of Hamiltonian form, with

$$
H(p, q) = T(p) + V(q) - \tfrac{1}{2}hT'(p)V'(q) \\
+ \tfrac{1}{12}h^2\big(V''(q)T'(p)^2 + T''(p)V'(q)^2\big). \tag{15.24}
$$

This example illustrates a phenomenon that holds true in great generality: numerical methods that give symplectic maps on Hamiltonian ODEs give rise to modified equations of any desired order that are themselves Hamiltonian. So these methods are extremely well described by problems that are (a) close to the original ODE and (b) have the same structure as the original ODE. By contrast, nonsymplectic methods, such as the Euler method that we examined on the pendulum problem, do not have this desirable property.

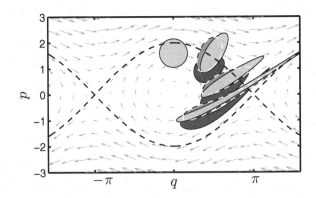

Fig. 15.8 Area evolution: Light shading for symplectic Euler method applied to the pendulum system (15.6), medium shading with white dashed border for the modified equation (15.20), dark shading for the exact flow map

15.5 Discussion

In 1988 three researchers, Lasagni, Sanz-Serna, and Suris, independently published a condition that characterizes symplectiness for Runge–Kutta methods. This turned out to be the same as the condition (14.25) that characterizes preservation of quadratic invariants. Many other authors have looked at designing and analysing numerical methods for general Hamiltonian ODEs or for specific classes arising in particular applications. The modified equation viewpoint appears to offer the best set of analytical tools for quantifying the benefit of symplecticness, and many interesting issues can be looked at. For example, the Kepler problem (14.15) is not only of Hamiltonian form with a quadratic invariant, but also has a property known as *reversibility*. Which combination of these three properties is it possible, or desirable, for a numerical method to match? In addition to symplecticness, there are many other closely related problems involving *Lie group* structures, and these ideas also extend readily beyond ODEs into the realms of partial and stochastic differential equations. Overall, there are many open questions in the field of geometric integration, and the area remains very active.

EXERCISES

15.1.** The product Jv corresponds to a clockwise rotation of v through a right angle. Deduce from the formula for the scalar product of two vectors that $u^T Jv = \|u\| \|v\| \sin\theta$, where θ is the angle between u and v and $\|u\| = (u^T u)^{1/2}$. Hence prove that the oriented area is given by the expression (15.1).

15.2.* Let $A(\alpha)$ denote the rotation matrix in Exercise 7.8. Show that $\mathrm{area}_o(A(\alpha)u, A(\alpha)v) = \mathrm{area}_o(u, v)$ for any two vectors u, $v \in \mathbb{R}^2$ and any angle α. This proves that the oriented area is unaffected when both vectors are rotated through equal angles.

15.3.* Show that the oriented area function (15.1) is linear, in the sense that $\mathrm{area}_o(x + z, y) = \mathrm{area}_o(x, y) + \mathrm{area}_o(z, y)$. Draw a picture to illustrate this result. (Hint: from Exercise 15.2 it is OK to assume that y is parallel with the vertical axis.)

15.4.* Let
$$B = \begin{bmatrix} a & c \\ b & d \end{bmatrix} \in \mathbb{R}^{2\times2}.$$

Show that the vertices of the parallelogram in Figure 15.1 arise when B is applied to the vertices of the unit square. Also, show that the oriented area has the form $\det(B)$.

15.5.* For $A \in \mathbb{R}^{2 \times 2}$, show that the left-hand side of (15.2) is equivalent to the matrix

$$\begin{bmatrix} 0 & \det(A) \\ -\det(A) & 0 \end{bmatrix}.$$

This confirms that the condition (15.2) is equivalent to $\det(A) = 1$.

15.6.* Show that the *Cremona* map [30]

$$\begin{bmatrix} x_1 \\ x_2 \end{bmatrix} \mapsto \begin{bmatrix} x_1 \cos \lambda - (x_2 - x_1^2) \sin \lambda \\ x_1 \sin \lambda + (x_2 - x_1^2) \cos \lambda \end{bmatrix},$$

where λ is constant, is symplectic.

15.7.** The backward Euler method applied to the pendulum ODE (15.6) corresponds to the implicit relations

$$\begin{bmatrix} p_{n+1} \\ q_{n+1} \end{bmatrix} = \begin{bmatrix} p_n - h \sin q_{n+1} \\ q_n + h p_{n+1} \end{bmatrix}. \tag{15.25}$$

Taking partial derivatives, using the chain rule, show that

$$\begin{bmatrix} 1 & h \cos q_{n+1} \\ -h & 1 \end{bmatrix} \begin{bmatrix} \dfrac{\partial p_{n+1}}{\partial p_n} & \dfrac{\partial p_{n+1}}{\partial q_n} \\ \dfrac{\partial q_{n+1}}{\partial p_n} & \dfrac{\partial q_{n+1}}{\partial q_n} \end{bmatrix} = \begin{bmatrix} 1 & 0 \\ 0 & 1 \end{bmatrix}.$$

Deduce that

$$\det \left(\begin{bmatrix} \dfrac{\partial p_{n+1}}{\partial p_n} & \dfrac{\partial p_{n+1}}{\partial q_n} \\ \dfrac{\partial q_{n+1}}{\partial p_n} & \dfrac{\partial q_{n+1}}{\partial q_n} \end{bmatrix} \right) = \frac{1}{1 + h^2 \cos q_{n+1}}.$$

This shows that the backward Euler method does not produce a symplectic map on this problem for any step size $h > 0$.

15.8.* With $H(p, q)$ defined by (15.7), show that the separatrices $H(p, q) = 1$ (shown as dashed lines in Figure 15.4) are given by $p = \pm 2 \cos(q/2)$.

15.9.*** The "adjoint" of the symplectic Euler method (15.11) has the form[2]

$$p_{n+1} = p_n - h \frac{\partial H}{\partial q}(p_n, q_{n+1}),$$

$$\tag{15.26}$$

$$q_{n+1} = q_n + h \frac{\partial H}{\partial p}(p_n, q_{n+1}).$$

---

[2]Some authors refer to (15.26) as the symplectic Euler method, in which case (15.11) becomes the adjoint.

Show that this method is also symplectic. Can it be implemented as an explicit method on separable problems of the form (15.12)?

15.10.*** Show that the implicit mid-point rule (14.13) is symplectic.

15.11.* Show that the symplectic Euler method (15.13) reduces to the method used in Example 13.4 when $V(q) = \frac{1}{2}q^2$ and $T(p) = \frac{1}{2}p^2$. Verify also that the Hamiltonian (15.21) reduces to a multiple of (13.20).

15.12.*** (Based on Sanz-Serna and Calvo [61, Example 10.1] and Beyn [3].) The time t flow map of a constant coefficient linear system $\boldsymbol{x}'(t) = A\boldsymbol{x}(t)$, where $A \in \mathbb{R}^{m \times m}$, may be written

$$\psi_t(\boldsymbol{x}_0) = \exp(At)\,\boldsymbol{x}_0.$$

Here, exp denotes the *matrix exponential*, which may be defined by extending the usual scalar power series to the matrix analogue:

$$\exp(X) = I + X + \frac{1}{2!}X^2 + \frac{1}{3!}X^3 + \cdots.$$

Show that the time h flow map of the $m = 2$ system (which depends on h)

$$\begin{bmatrix} u'(t) \\ v'(t) \end{bmatrix} = \left(h^{-1} \log \begin{bmatrix} 1 & -h \\ h & 1-h^2 \end{bmatrix} \right) \begin{bmatrix} u(t) \\ v(t) \end{bmatrix}$$

corresponds to one step of the symplectic Euler method (15.11) on the Hamiltonian ODE with $H(p,q) = (p^2 + q^2)/2$. Here, log denotes the *matrix logarithm* [37], for which $\exp(\log(A)) = A$. This is one of the rare cases where we are able to find an exact modified equation, rather than truncate at some point in an h-expansion. Find an equivalent exact modified equation for the adjoint method (15.26) applied to the same Hamiltonian problem.

15.13.*** By following the derivation of (15.20), construct a modified equation for the adjoint of the symplectic Euler method, defined in (15.26), when applied to a separable problem of the form (15.14). Also, show that this modified equation is Hamiltonian.

15.14.*** By matching expansions up to $\mathcal{O}(h^4)$, derive a modified equation of the form (15.23) for the symplectic Euler method (15.13) applied to the separable Hamiltonian problem (15.14), and show that the result is a Hamiltonian problem with $H(p,q)$ as in (15.24).

Stochastic Differential Equations

16.1 Introduction

Many mathematical modelling scenarios involve an inherent level of *uncertainty*. For example, rate constants in a chemical reaction model might be obtained experimentally, in which case they are subject to measurement errors. Or the simulation of an epidemic might require an educated guess for the initial number of infected individuals. More fundamentally, there may be microscopic effects that (a) we are not able or willing to account for directly, but (b) can be approximated stochastically. For example, the dynamics of a coin toss could, in principle, be simulated to high precision if we were prepared to measure initial conditions sufficiently accurately and take account of environmental effects, such as wind speed and air pressure. However, for most practical purposes it is perfectly adequate, and much more straightforward, to model the outcome of the coin toss using a random variable that is equally likely to take the value heads or tails. Stochastic models may also be used in an attempt to deal with ignorance. For example, in mathematical finance, there appears to be no universal "law of motion" for the movement of stock prices, but random models seem to fit well to real data.

There are many ways to incorporate randomness into quantitative modelling and simulation. We focus here, very briefly, on a specific approach that is mathematically elegant and is becoming increasingly popular. The underlying theory is relatively new in comparison with Newton's calculus—the classic work by Ito was done in the 1940s, and the design and study of numerical methods is

D.F. Griffiths, D.J. Higham, *Numerical Methods for Ordinary Differential Equations*,
Springer Undergraduate Mathematics Series, DOI 10.1007/978-0-85729-148-6_16,
© Springer-Verlag London Limited 2010

an active research topic. Our aim here is to give a nontechnical and accessible introduction, in the manner of the expository article by Higham [31]. For further information, in roughly increasing order of technical difficulty, we recommend Mikosch [51], Cyganowski et al. [13], Mao [49], Milsein and Tretyakov [52] and Kloeden and Platen [42].

16.2 Random Variables

In this chapter, we deal informally with scalar continuous *random variables*, which take values in the range $(-\infty, \infty)$. We may characterize such a random variable, X, by the probability that X lies in any interval $[a, b]$:

$$\mathbb{P}\left(a \leq X \leq b\right) = \int_a^b p(y)\, \mathrm{d}y. \tag{16.1}$$

Here, p is known as the *probability density function* for X, and the left-hand side of (16.1) is read as "the probability that X lies between a and b." Figure 16.1 illustrates this idea: the chance of X taking values between a and b is given by the corresponding area under the curve. So regions where the density p is large correspond to likely values. Because probabilities cannot be negative, and because X must lie somewhere, a density function p must satisfy

(a) $p(y) \geq 0$, for all $y \in \mathbb{R}$;

(b) $\displaystyle\int_{-\infty}^{\infty} p(y)\, \mathrm{d}y = 1$.

An extremely important case is

$$p(y) = \frac{1}{\sqrt{2\sigma^2\pi}} \exp\left(-\frac{(y-\mu)^2}{2\sigma^2}\right), \tag{16.2}$$

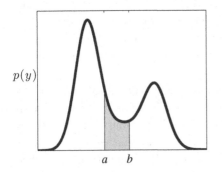

$p(y)$

$a \quad b$

Fig. 16.1 Illustration of the identity (16.1). The probability that X lies between a and b is given by the area under the probability density function for $a \leq y \leq b$

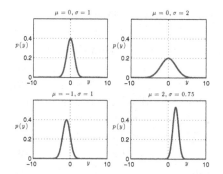

Fig. 16.2 Some density functions (16.2) corresponding to normally distributed random variables

where μ and $\sigma \geq 0$ are fixed parameters. Examples with $\mu = 0$ and $\sigma = 1$, $\mu = 0$ and $\sigma = 2$, $\mu = -1$ and $\sigma = 1$, and $\mu = 2$ and $\sigma = 0.75$ are plotted in Figure 16.2. If X has a probability density function given by (16.2) then we write $X \sim N(\mu, \sigma^2)$ and say that X is a *normally distributed* random variable.[1] In particular, when $X \sim N(0, 1)$ we say that X has the *standard normal distribution*. The bell-shaped nature of p in (16.2) turns out to be ubiquitous—the celebrated *Central Limit Theorem* says that, loosely, whenever a large number of random sources are combined *whatever their individual nature, the overall effect can be well approximated by a single normally distributed random variable.*

The parameters μ and σ in (16.2) play well-defined roles. The value $y = \mu$ corresponds to a peak and an axis of symmetry of p, and σ controls the "spread." More generally, given a random variable X with probability density function p, we define the *mean*, or *expected value*, to be[2]

$$\mathbb{E}[X] := \int_{-\infty}^{\infty} y \, p(y) \, \mathrm{d}y. \tag{16.3}$$

It is easy to check that $\mathbb{E}[X] = \mu$ with the normal density (16.2); see Exercise 16.1. The *variance* of X may then be defined as the expected value of the new random variable $(X - \mathbb{E}[X])^2$; that is,

$$\mathsf{var}[X] := \mathbb{E}\left[(X - \mathbb{E}[X])^2\right]. \tag{16.4}$$

Intuitively, the size of the variance captures the extent to which X may vary about its mean. In the normal case (16.2) we have $\mathsf{var}[X] = \sigma^2$; see Exercise 16.3.

---

[1]The word normal here does not imply "common" or "typical"; it stems from the geometric concept of orthogonality.

[2]In this informal treatment, any integral that we write is implicitly assumed to take a finite value.

16.3 Computing with Random Variables

A *pseudo-random number generator* is a computational algorithm that gives us a new number, or sample, each time we request one. En masse, these samples appear to agree with a specified density function. In other words, recalling the picture in Figure 16.1, if we call the pseudo-random number generator lots of times, then the proportion of samples in any interval $[a, b]$ would be approximately the same as the integral in (16.1). If we think of each sample coming out of the pseudo-random number generator as being the result of an independent trial, then we can make the intuitively reasonable connection between the *probability* of the event $a \leq X \leq b$ and the *frequency* at which that event is observed over a long sequence of independent trials. For example, if a and b are chosen such that $\mathbb{P}(a \leq X \leq b) = \frac{1}{2}$, then, in the long term, we would expect half of the calls to a suitable pseudo-random number generator to produce samples lying in the interval $[a, b]$.

In this book we will assume that a pseudo-random number generator corresponding to a random variable $X \sim \mathsf{N}(0, 1)$ is available. For example, MATLAB has a built-in function **randn**, and 10 calls to this function produced the samples

```
-0.4326
-1.6656
 0.1253
 0.2877
-1.1465
 1.1909
 1.1892
-0.0376
 0.3273
 0.1746
```

and another ten produced

```
-0.1867
 0.7258
-0.5883
 2.1832
-0.1364
 0.1139
 1.0668
 0.0593
-0.0956
-0.8323
```

Suppose we generate M samples from **randn**. We may divide the y-axis into bins of length Δ_y and let N_i denote the number of samples in each subinterval $[i\Delta_y, (i+1)\Delta_y]$. Then the integral-based definition of probability (16.1) tells us that

$$\mathbb{P}\left(i\Delta_y \leq X \leq (i+1)\Delta_y\right) = \int_{i\Delta_y}^{(i+1)\Delta_y} p(y)\,dy \approx p(i\Delta_y)\Delta_y. \qquad (16.5)$$

This was obtained by approximating the area under the curve by the area of a rectangle with height $p(i\Delta_y)$ and base Δ_y; this is valid when Δ_y is small. On the other hand, identifying the probability of an event with its long-term observed frequency, we have

$$\mathbb{P}\left(i\Delta_y \leq X \leq (i+1)\Delta_y\right) \approx \frac{N_i}{M}. \qquad (16.6)$$

Combining (16.5) and (16.6) gives us

$$p(i\Delta_y) \approx \frac{N_i}{\Delta_y M}. \qquad (16.7)$$

Figure 16.3 makes this concrete. Here, we took $M = 10000$ samples from MATLAB's **randn** and used 17 subintervals on the y-axis, with centres at -4, -3.5, -3, ..., 3.5, 4. The histogram shows the appropriately scaled proportion of samples lying in each interval; the height of each rectangle is given by the right-hand side of (16.7). We have superimposed the probability density function (16.2) with $\mu = 0$ and $\sigma = 1$, which is seen to match the computed data.

In many circumstances, our aim is to find the expected value of some random variable, X, and we are able to use a pseudo-random number generator to compute samples from its distribution. If $\{\xi_i\}_{i=1}^M$ is a set of such samples, then it is intuitively reasonable that the *sample mean*

$$a_M := \frac{1}{M}\sum_{i=1}^M \xi_i \qquad (16.8)$$

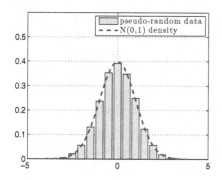

Fig. 16.3 Histogram of samples from a $N(0,1)$ pseudo-random number generator, with probability density function (16.2) for $\mu = 0$ and $\sigma = 1$ superimposed. This illustrates the idea that en masse the samples appear to come from the appropriate distribution

can be used to approximate $\mathbb{E}[X]$. Standard statistical results[3] can be used to show that in the asymptotic $M \to \infty$ limit, the range

$$\left[a_M - \frac{1.96\sqrt{\mathrm{var}[X]}}{\sqrt{M}}, a_M + \frac{1.96\sqrt{\mathrm{var}[X]}}{\sqrt{M}} \right] \qquad (16.9)$$

is a *95% confidence interval* for $\mathbb{E}[X]$. This may be understood as follows: if we were to repeat the computation of a_M in (16.8) many times, each time using fresh samples from our pseudo-random number generator, then the statement "the exact mean lies in this interval" would be true 95% of the time. In practice, we typically do not have access to the exact variance, $\mathrm{var}[X]$, which is required in (16.9). From (16.4) we see that the variance is itself a particular case of an expected value, so the idea in (16.8) can be repeated to give the *sample variance*

$$b_M^2 := \frac{1}{M} \sum_{i=1}^{M} (\xi_i - a_M)^2 .$$

Here, the unknown expected value $\mathbb{E}[X]$ has been replaced by the sample mean a_M.[4] Hence, instead of (16.9) we may use the more practical alternative

$$\left[a_M - \frac{1.96\sqrt{b_M^2}}{\sqrt{M}}, a_M + \frac{1.96\sqrt{b_M^2}}{\sqrt{M}} \right]. \qquad (16.10)$$

As an illustration, consider a random variable of the form

$$X = \mathrm{e}^{-1+2Y}, \text{ where } Y \sim \mathsf{N}(0,1).$$

In this case, we can compute samples by calling a standard normal pseudo-random number generator, scaling the output by 2 and shifting by -1 and then exponentiating. Table 16.1 shows the sample means (16.8) and confidence intervals (16.10) that arose when we used $M = 10^2, 10^3, \ldots, 10^7$. For this simple example, it can be shown that the exact mean has the value $\mathbb{E}[X] = 1$ (see Exercise 16.4) so we can judge the accuracy of the results. Of course, this type of computation, which is known as a *Monte Carlo simulation*, is useful in those circumstances where the exact mean cannot be obtained analytically. We see that the accuracy of the sample mean and the precision of the confidence interval improve as the number of samples, M, increases. In fact, we can see directly from the definition (16.10) that the width of the confidence interval scales with M like $1/\sqrt{M}$; so, *to obtain one more decimal place of accuracy we need to do 100 times more computation*. For this reason, Monte Carlo simulation is impractical when very high accuracy is required.

---

[3]More precisely, the Strong Law of Large Numbers and the Central Limit Theorem.

[4]There is a well-defined sense in which this version of the sample variance is improved if we multiply it by the factor $M/(M-1)$. However, a justification for this is beyond the scope of the book, and the effect is negligible when M is large.

M	a_M	confidence interval
10^2	0.9445	$[0.4234, 1.4656]$
10^3	0.8713	$[0.6690, 1.0737]$
10^4	1.0982	$[0.9593, 1.2371]$
10^5	1.0163	$[0.9640, 1.0685]$
10^6	0.9941	$[0.9808, 1.0074]$
10^7	1.0015	$[0.9964, 1.0066]$

Table 16.1 Sample means and confidence intervals for Monte Carlo simulationsof a random variable X for which $\mathbb{E}[X] = 1$

16.4 Stochastic Differential Equations

We know that the Euler iteration

$$x_{n+1} = x_n + hf(x_n)$$

produces a sequence $\{x_n\}$ that converges to a solution of the ODE $x'(t) = f(x(t))$. To introduce a stochastic element we will give each x_{n+1} a random "kick," so that

$$x_{n+1} = x_n + hf(x_n) + \sqrt{h}\,\xi_n\,g(x_n). \qquad (16.11)$$

Here, ξ_n denotes the result of a call to a standard normal pseudo-random number generator and g is a given function. So the size of the random kick is generally *state-dependent*—it depends upon the current approximation x_n, via the value $g(x_n)$. We also see a factor $\sqrt{h}$ in the random term. Why is it not h, or $h^{1/4}$ or h^2? It turns out that $\sqrt{h}$ is the correct scaling when we consider the limit $h \to 0$. A larger power of h would cause the noise to disappear and a smaller power of h would cause the noise to swamp out the original ODE completely.

Given functions f and g and an initial condition $x(0)$, we can think of a process $x(t)$ that arises when we take the $h \to 0$ limit in (16.11). In other words, just as in the deterministic case, we can fix t and consider the limit as $h \to 0$ of x_N where $Nh = t$. Of course, this construction for $x(t)$ leads to a random variable—each set of pseudo-random numbers $\{\xi_n\}_{n=0}^{N-1}$ gives us a new sample from the distribution of $x(t)$.

Assuming that this $h \to 0$ limit is valid, we will refer to $x(t)$ as the solution to a *stochastic differential equation* (SDE). There are thus three ingredients for an SDE

– A function f, which plays the same role as the right-hand side of an ODE. In the SDE context this is called the *drift coefficient*.

- A function g, which affects the size of the noise contribution. This is known as the *diffusion coefficient.*

- An initial condition, $x(0)$. The initial condition might be deterministic, but more generally it is allowed to be a random variable—in that case we simply use a pseudo-random number generator to pick the starting value x_0 for (16.11).

As an illustration, we will consider the case where f and g are linear; that is,

$$f(x) = ax, \qquad g(x) = bx, \tag{16.12}$$

where a and $b > 0$ are constants, and we fix $x(0) = 1$. For the dark curve in Figure 16.4 we take $a = 2$, $b = 1$, and $x(0) = 1$, and show the results of applying the iteration (16.11) with a step size $h = 10^{-3}$ for $0 \leq t \leq 1$. The closely spaced, but discrete, points $\{x_n\}$ have been joined by straight lines for clarity. This gives the impression of a continuous but jagged curve. This can be formalized—the limiting $h \to 0$ solution produces paths that are continuous but nowhere differentiable. For the lighter curve in Figure 16.4 we repeated the experiment with different pseudo-random samples ξ_n and with the diffusion strength b increased to 2. We see that this gives a more noisy or "volatile" path.

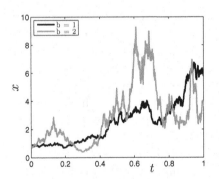

Fig. 16.4 Results for the iteration (16.11) with drift and diffusion coefficients from (16.12) and a step size $h = 10^{-3}$. Here, for the same drift strength, $a = 2$, we show a path with $b = 1$ (dark) and $b = 2$ (light)

We emphasize that the details in Figure 16.4 would change if we repeated the experiments with fresh pseudo-random numbers. To illustrate this idea, in the upper picture of Figure 16.5 we show 50 different paths, each computed as in Figure 16.4, with $a = 0.06$ and $b = 0.75$. At the final time, $t = 1$, each path produces a single number that, in the $h \to 0$ limit, may be regarded as a sample from the distribution of the random variable $x(1)$ describing the SDE solution at $t = 1$. In the lower picture of Figure 16.5 we produced a histogram for 10^4 such samples. Overall, the figure illustrates two different ways to think about an SDE. We can consider individual paths evolving over time, as in the upper picture, or we can fix a point in time and consider the distribution of values at that point, as in the lower picture. From the latter perspective, by studying this simple SDE analytically it can be shown, given a deterministic initial condition, that for the exact solution the random variable $x(t)$ at time t has a so-called *lognormal* probability density function given by

$$p(y) = \frac{\exp\left(\frac{-[\log(y/x(0))-(a-\frac{1}{2}b^2)t]^2}{2b^2t}\right)}{yb\sqrt{2\pi t}}, \qquad \text{for } y > 0, \qquad (16.13)$$

and $p(y) = 0$ for $y \leq 0$. We have superimposed this density function for $t = 1$ in the lower picture of Figure 16.5, and we see that it matches the histogram closely. Exercise 16.6 asks you to check that (16.13) defines a valid density function, and to confirm that the mean and variance take the form

$$\mathbb{E}[x(t)] = x_0\, e^{at}, \qquad (16.14)$$

$$\text{var}[x(t)] = x_0^2\, e^{2at}\left(e^{b^2t} - 1\right). \qquad (16.15)$$

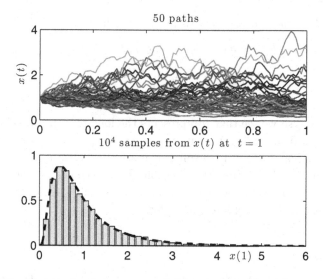

Fig. 16.5 Upper: In the manner of Figure 16.4, 50 paths using (16.11) for the linear case (16.12) with $a = 0.06$ and $b = 0.75$. Lower: In the manner of Figure 16.3, binned path values at the final time $t = 1$ and probability density function (16.13) superimposed as a dashed curve.

16.5 Examples of SDEs

In this section we mention a few examples of SDE models that have been proposed in various application areas.

Example 16.1

In mathematical finance, SDEs are often used to represent quantities whose future values are uncertain. The linear case (16.12) is by far the most popular model for *assets*, such as company share prices, and it forms the basis of the celebrated Black–Scholes theory of option valuation [32]. If $b = 0$ then we revert to the simple deterministic ODE $x'(t) = ax(t)$, for which $x(t) = e^{at}x(0)$. This describes the growth of an investment that has a guaranteed fixed rate a. The stochastic term that arises when $b \neq 0$ reflects the uncertainty in the rate of return for a typical financial asset.[5] In this context, b is called the *volatility*.

Example 16.2

A *mean-reverting square-root process* is an SDE with

$$f(x) = a\left(\mu - x\right), \qquad g(x) = b\sqrt{x}, \tag{16.16}$$

where a, μ, and $b > 0$ are constants. For this SDE

$$\mathbb{E}[x(t)] = \mu + e^{-at}(\mathbb{E}[X(0)] - \mu) \tag{16.17}$$

(see Exercise 16.9) so we see that μ represents the long-term mean value. This model is often used to represent an interest rate; and in this context it is associated with the names Cox, Ingersoll and Ross [11]. Given a positive initial condition, it can be shown that the solution never becomes negative, so the square root in the diffusion coefficient, $g(x)$, always makes sense. See, for example, Kwok [43] and Mao [49] for more details.

Example 16.3

The traditional logistic ODE for population growth can be generalized to allow for stochastic effects by taking

$$f(x) = rx\left(K - x\right), \qquad g(x) = \beta x, \tag{16.18}$$

where r, K and $\beta > 0$ are constants. Here, $x(t)$ denotes the population density of a species at time t, with carrying capacity K and characteristic timescale $1/r$ [56], and β governs the strength of the environmental noise.

---

[5]Or the authors' pension funds.

Example 16.4

The case where

$$f(x) = r\,(G - x), \qquad g(x) = \sqrt{\varepsilon x\,(1 - x)}, \tag{16.19}$$

with $r, \varepsilon > 0$ and $0 < G < 1$ is suggested by Cobb [9] as a model for the motion over time of an individual through the liberal–conservative political spectrum. Here $x(t) = 0$ denotes an extreme liberal and $x(t) = 1$ denotes an extreme conservative. With this choice of diffusion term, $g(x)$, a person with extreme views is less likely to undergo random fluctuations than one nearer the centre of the political spectrum.

Example 16.5

The SDE with

$$f(x) = -\mu\left(\frac{x}{1 - x^2}\right), \qquad g(x) = \sigma, \tag{16.20}$$

with μ and $\sigma > 0$, is proposed by Lesmono et al. [46]. Here, in an environment where two political parties, A and B, are dominant, $x(t)$ represents the difference $P_A(t) - P_B(t)$, where $P_A(t)$ and $P_B(t)$ denote the proportions of the population that intend to vote for parties A and B, respectively, at time t.

Example 16.6

Letting $V(x)$ denote the *double-well potential*

$$V(x) = x^2(x - 2)^2; \tag{16.21}$$

as shown in Figure 16.6, we may construct the SDE with

$$f(x) = -V'(x), \qquad g(x) = \sigma. \tag{16.22}$$

When $\sigma = 0$, we see from the chain rule that the resulting ODE $x'(t) = -V'(x(t))$ satisfies

$$\frac{\mathrm{d}}{\mathrm{d}t}V(x(t)) = V'(x(t))\frac{\mathrm{d}}{\mathrm{d}t}x(t) = -\left(V'(x(t))\right)^2.$$

So, along any solution curve, $x(t)$, the potential $V(\cdot)$ is nonincreasing. Moreover, it strictly decreases until it reaches a stationary point; with reference to Figure 16.6, any solution with $x(0) \neq 1$ slides down a wall of the potential well and comes to rest at the appropriate minimum $x = 0$ or $x = 2$. In the additive noise case, $\sigma > 0$, it is now possible for a solution to overcome the *potential barrier* that separates the two stable rest states—a path may occasionally jump over the central hump and thereby move from the vicinity of one well to the other. This gives a simple caricature of a *bistable switching* mechanism that is important in biology and physics.

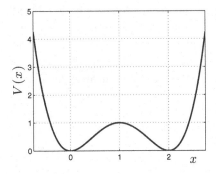

Fig. 16.6 Double–well potential function $V(x)$ from (16.21)

16.6 Convergence of a Numerical Method

Our informal approach here is to regard an SDE as whatever arises when we take the $h \rightarrow 0$ limit in the iteration (16.11). For any fixed $h > 0$, we may then interpret (16.11) as a numerical method that allows us to compute approximations for this SDE. In fact, this way of extending the basic Euler method gives what is known as the *Euler–Maruyama* method. If we focus on the final-time value, t_f, then we may ask how accurately the numerical method can approximate the random variable $x(t_f)$ from the SDE. For the example in Figure 16.5, the lower picture shows that the histogram closely matches the correct density function. Letting $t_f = nh$, so that $n \rightarrow \infty$ and $h \rightarrow 0$ with t_f fixed, how can we generalize the concept of order of convergence that we developed for ODEs?

It turns out that there are many, nonequivalent, ways in which to measure the accuracy of the numerical method. If we let x_n denote the random variable corresponding to the numerical approximation at time t_f, so that the endpoint of each path in the upper picture of Figure 16.5 gives us a sample for x_n, then we could study the difference between the expected values of the random variables $x(t_f)$ and x_n. This quantifies what is known as the *weak error* of the method,[6] and it can be shown that this error decays at first order; that is,

$$\mathbb{E}\left[x(t_f)\right] - \mathbb{E}\left[x_n\right] = \mathcal{O}(h). \tag{16.23}$$

---

[6]More generally, the weak error can be defined by comparing *moments*—expected values of powers of $x(t_f)$ and x_n.

In Figure 16.7 we compute approximations to the weak error, as defined in the left-hand side of (16.23), for the linear SDE given by (16.12) with $a = 2$, $b = 1$ and $x(0) = 1$. The asterisks show the weak error values for a range of step sizes h. In each case, we used the known exact value (16.14) for the mean of the SDE solution, and to approximate the mean of the numerical method we computed a large number of paths and used the sample mean (16.8), making sure that the 95% confidence intervals were negligible relative to the actual weak errors. Figure 16.7 uses a log-log scale, and the asterisks appear to lie roughly on a straight line. A reference line with slope equal to one is shown. In the least-squares sense, the best straight line approximation to the asterisk data gives a slope of 0.9858 with a residual of 0.0508. So, overall, the weak errors are consistent with the first-order behaviour quoted in (16.23).

The first-order rate of weak convergence in (16.23) matches what we know about the deterministic case—when we switch off the noise, $g \equiv 0$, the method reverts to standard Euler, for which convergence of order 1 is attained.

On the other hand, if we are not concerned with "the error of the means" but rather "the mean of the error," then it may be more appropriate to look at the *strong error* $\mathbb{E}[|x(t_f) - x_n|]$. It can be shown that this version of the error decays at a rate of only one half; that is,

$$\mathbb{E}\left[|x(t_f) - x_n|\right] = \mathcal{O}(h^{1/2}). \tag{16.24}$$

To make this concrete, for the same SDE and step sizes as in the weak tests, the circles in Figure 16.7 show approximations to the strong error. Samples for the "exact" SDE solution, x_n, were obtained by following more highly resolved paths—see, for example, Higham [31] for more information. The circles in the figure appear to lie approximately on a straight line that agrees with the reference line of slope one half, and a least-squares fit to the circle data gives a slope of 0.5384 with residual 0.0266, consistent with (16.24).

We emphasize that this result marks a significant departure from the deterministic ODE case, where the underlying Euler method attains first order. The

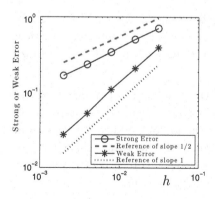

Fig. 16.7 Asterisks show weak errors (16.23) and circles show strong errors (16.24) for the Euler–Maruyama method (16.11) applied to the linear SDE (16.12)

degradation in approximation power when we measure error in the strong sense of (16.24) is caused by a lack of smoothness that is evident from Figures 16.4 and 16.5; we cannot appeal directly to the type of Taylor series expansion that served us so well in the preceding chapters, and a complete understanding requires the tools of stochastic calculus.

16.7 Discussion

Although we have focussed here on scalar problems, the concept of SDEs extends naturally to the case of systems, $x(t) \in \mathbb{R}^m$, with multiple noise terms, and the Euler–Maruyama method (16.11) carries through straightforwardly, retaining the same weak and strong order.

For deterministic ODEs, Euler's method would typically be dismissed as having an order of convergence that is too low to be of practical use. For SDEs, however, Euler–Maruyama, and certain low order implicit variations, are widely used in practice. There are two good reasons.

1. Designing higher order methods is much more tricky, especially in the case of systems, and those that have been put forward tend to have severe computational overheads.

2. Souping-up the iteration (16.11) is of limited use if the simulations are part of a Monte Carlo style computation. In that case the relatively slow $O(1/\sqrt{M})$ rate at which the confidence interval shrinks is likely to provide the bottleneck.

Following up on point 2, we mention that recent work of Giles [21] makes it clear that, in a Monte Carlo context, where we wish to compute the expected value of a quantity involving the solution of an SDE,

– the concepts of weak and strong convergence are both useful, and

– it is worthwhile to study the interaction between the weak and strong discretization errors arising from the timestepping method and the statistical sampling error arising from the Monte Carlo method, in order to optimize the overall efficiency.

Researchers in numerical methods for SDEs are actively pursuing many of the issues that we have discussed in the ODE context, such as analysis of stability, preservation of geometric features, construction of modified equations and design of adaptive step size algorithms. There are also many new challenges, including the development of *multiscale* methods that mix together deterministic and stochastic regimes in order to model complex systems.

EXERCISES

16.1.* Using the definition of the mean (16.3), show that $\mathbb{E}[X] = \mu$ for $X \sim \mathsf{N}(\mu, \sigma^2)$. [Hint: you may use without proof the fact that $\int_{-\infty}^{\infty} p(y)\, dy = 1$ for $p(y)$ in (16.2).]

16.2.* Given that the expectation operator is *linear*, so that for any two random variables X and Y and any $\alpha, \beta \in \mathbb{R}$,

$$\mathbb{E}[\alpha X + \beta Y] = \alpha \mathbb{E}[X] + \beta \mathbb{E}[Y], \qquad (16.25)$$

show that the variance, defined in (16.4), may also be written

$$\mathsf{var}[X] := \mathbb{E}\left[X^2\right] - (\mathbb{E}[X])^2. \qquad (16.26)$$

16.3.** If the random variable X has probability density function $p(y)$ then, generally, for any function h, we have

$$\mathbb{E}[h(X)] := \int_{-\infty}^{\infty} h(y)\, p(y)\, dy. \qquad (16.27)$$

Following on from Exercises 16.1 and 16.2 show that for $X \sim \mathsf{N}(\mu, \sigma^2)$ we have $\mathbb{E}[X^2] = \mu^2 + \sigma^2$ and hence $\mathsf{var}[X] = \sigma^2$.

16.4.** Using $h(y) = \exp(-1 + 2y)$ in (16.27), show that $\mathbb{E}[X] = 1$ for $X = \exp(-1 + 2Y)$ with $Y \sim \mathsf{N}(0, 1)$.

16.5.* Based on the definition of the confidence interval (16.10), if we did not already know the exact answer, roughly how many more rows would be needed in Table 16.1 in order for us to be 95% confident that we have correctly computed the first five significant digits in the expected value?

16.6.** Show that $p(y)$ in (16.13) satisfies $\int_{-\infty}^{\infty} p(y)\, dy = 1$ and, by evaluating $\int_{-\infty}^{\infty} y\, p(y)\, dy$ and $\int_{-\infty}^{\infty} y^2\, p(y)\, dy$, confirm the expressions (16.14) and (16.15).

16.7.*** For the case where $f(x) = ax$ and $g(x) = bx$, the Euler–Maruyama method (16.11) takes the form

$$x_{k+1} = (1 + ha)\, x_k + \sqrt{h}\, b Z_k x_k, \qquad (16.28)$$

where each $Z_k \sim \mathsf{N}(0, 1)$. Suppose the initial condition $x(0) = x_0$ is deterministic. By construction, we have $\mathbb{E}[Z_k] = 0$ and $\mathbb{E}[Z_k^2] =$

$\text{var}[Z_k] = 1$. Because a fresh call to a pseudo-random number generator is made on each step, we can say that Z_k and x_k are *independent*, and it follows that

$$\mathbb{E}[x_k Z_k] = \mathbb{E}[x_k]\mathbb{E}[Z_k] = 0,$$
$$\mathbb{E}[x_k^2 Z_k] = \mathbb{E}[x_k^2]\mathbb{E}[Z_k] = 0,$$
$$\mathbb{E}[x_k^2 Z_k^2] = \mathbb{E}[x_k^2]\mathbb{E}[Z_k^2] = \mathbb{E}[x_k^2].$$

Taking expectations in (16.28), and using the linearity property (16.25), we find that

$$\begin{aligned}
\mathbb{E}[x_{k+1}] &= \mathbb{E}\left[x_k(1+ha) + \sqrt{h}bx_k Z_k\right] \\
&= (1+ha)\,\mathbb{E}[x_k] + \sqrt{h}b\mathbb{E}[x_k Z_k] \\
&= (1+ha)\,\mathbb{E}[x_k].
\end{aligned}$$

So,

$$\mathbb{E}[x_n] = (1+ha)^n x_0.$$

Consider now the limit where $h \to 0$ and $n \to \infty$ with $nh = t_f$ fixed, as in the convergence analysis of Theorem 2.4. Show that

$$\mathbb{E}[x_n] \to e^{at_f}x_0, \qquad (16.29)$$

in agreement with the expression (16.14) for the SDE.

Similarly, squaring both sides in (16.28) and then taking expected values, and using the linearity property (16.25), we have

$$\begin{aligned}
\mathbb{E}[x_{k+1}^2] &= \mathbb{E}\left[x_k^2(1+ha)^2 + 2(1+ha)\sqrt{h}bx_k^2 Z_k + hb^2 x_k^2 Z_k^2\right] \\
&= (1+ha)^2\,\mathbb{E}[x_k^2] + 2(1+ha)\sqrt{h}b\mathbb{E}[x_k^2 Z_k] + hb^2\mathbb{E}[x_k^2 Z_k^2] \\
&= (1+ha)^2\,\mathbb{E}[x_k^2] + 0 + hb^2\mathbb{E}[x_k^2] \\
&= \left((1+ha)^2 + hb^2\right)\mathbb{E}[x_k^2].
\end{aligned}$$

By taking logarithms, or otherwise, show that in the same limit $h \to 0$ and $n \to \infty$ with $nh = t_f$ fixed we have

$$\mathbb{E}[x_n^2] \to e^{(2a+b^2)t_f}x_0^2,$$

so that

$$\text{var}[x_n] = e^{2at_f}\left(e^{b^2 t_f} - 1\right)x_0^2,$$

in agreement with the expression (16.15) for the SDE.

16.8.*** Repeat the steps in Exercise 16.7 to get expressions for the mean and variance in the *additive noise* case where $f(x) = ax$ and $g(x) = b$. This is an example of an *Ornstein–Uhlenbeck* process.

16.9.** Follow the arguments that led to (16.29) in order to justify the expression (16.17) for the mean-reverting square root process.

A
Glossary and Notation

AB: Adams–Bashforth—names of a family of explicit LMMs. AB(2) denotes the two-step, second-order Adams–Bashforth method.

AM: Adams–Moulton—names of a family of implicit LMMs. AM(2) denotes the two-step, third-order Adams–Moulton method.

BDF: backward differentiation formula.

BE: backward Euler (method).

C_{p+1} : error constant of a pth-order LMM.

CF: complementary function—the general solution of a homogeneous linear difference or differential equation.

$\triangle$E: difference equation.

$\mathbb{E}(\cdot)$: expected value.

f_n : the value of $f(t, x)$ at $t = t_n$ and $x = x_n$.

FE: forward Euler (method).

GE: global error—the difference between the exact solution $x(t_n)$ at $t = t_n$ and the numerical solution x_n: $e_n = x(t_n) - x_n$.

∇: gradient operator. $\nabla \mathcal{F}$ denotes the vector of partial derivatives of $\mathcal{F}$.

h: step size—numerical solutions are sought at times $t_n = t_0 + nh$ for $n = 0, 1, 2, \ldots$.

D.F. Griffiths, D.J. Higham, *Numerical Methods for Ordinary Differential Equations*,
Springer Undergraduate Mathematics Series, DOI 10.1007/978-0-85729-148-6,
© Springer-Verlag London Limited 2010

$\widehat{h}$: λh, where λ is the coefficient in the first-order equation $x'(t) = \lambda x(t)$ or an eigenvalue of the matrix A in the system of ODEs $\boldsymbol{x}'(t) = A\boldsymbol{x}(t)$.

I: identity matrix.

$\Im(\lambda)$: imaginary part of a complex number λ.

IAS: interval of absolute stability.

IC: initial condition.

IVP: initial value problem—an ODE together with initial condition(s).

$j = m : n$: for integers $m < n$ this is shorthand for the sequence of consecutive integers from m to n. That is, $j = m, m+1, m+2, \ldots, n$.

$\mathscr{L}_h$: linear difference operator.

LMM: linear multistep method.

LTE: local truncation error—generally the remainder term R in a Taylor series expansion.

ODE: ordinary differential equation.

$\mathbb{P}(a \leq X \leq b)$: probability that X lies in the interval $[a, b]$.

PS: particular solution—any solution of an inhomogeneous linear difference or differential equation.

$\rho(r)$: first characteristic polynomial of a LMM.

$\mathcal{R}$: region of absolute stability.

$\mathcal{R}_0$: interval of absolute stability.

$\Re(\lambda)$: real part of a complex number λ.

RK: Runge–Kutta. $RK(p)$ $(p \leq 4)$ is a p–stage, pth order RK method. $RK(p, q)$ denotes a pair of RK methods for use in adaptive time–stepping.

$R(\widehat{h})$: stability function of an RK method.

$\sigma(r)$: second characteristic polynomial of a LMM.

SDE: stochastic differential equation.

t_f: final time at which solution is required.

t_n: a grid point, generally, $t_n = t_0 + nh$, at which the numerical solution is computed.

T_n: local truncation error at time $t = t_n$.

$\widehat{T}_n$: In Chapter 11 it denotes an approximation to the local truncation error T_n, usually based on the leading term in its Taylor expansion. In Chapter 13 it denotes the local truncation error based on the solution of the modified, rather than the original, IVP.

TS: Taylor Series. TS(p)—the Taylor Series method with $p+1$ terms (up to, and including, pth derivatives).

var($\cdot$) : variance.

x: a scalar-valued quantity while $\boldsymbol{x}$ denotes a vector-valued quantity.

$x'(t)$: derivative of $x(t)$ with respect to t.

$x_n, x'_n, x''_n, \ldots$: approximations to $x(t_n), x'(t_n), x''(t_n), \ldots$.

$X \sim \mathsf{N}(\mu, \sigma^2)$: a normally distributed random variable with mean μ and variance σ^2.

B
Taylor Series

B.1 Expansions

The idea behind Taylor series expansions is to approximate a smooth function $g : \mathbb{R} \to \mathbb{R}$ locally by a polynomial. We suppose here that all derivatives of g are bounded. The approximating polynomial will be chosen to match as many derivatives of g as possible at a certain point, say a. Fixing a, we have, from the definition of a derivative,

$$\frac{g(a + h) - g(a)}{h} \approx g'(a), \tag{B.1}$$

for small h. This rearranges to

$$g(a + h) \approx g(a) + hg'(a). \tag{B.2}$$

As h varies, the right-hand side in (B.2) defines the tangent line to the function g based on the point $x = a$. This is a first order polynomial approximation, or Taylor series expansion with two terms—it is close to the function g for sufficiently small h. Differentiating with respect to h and setting $h = 0$ we see that the polynomial $g(a) + hg'(a)$ in (B.2) matches the zeroth and first derivatives of g. We may improve the accuracy of the approximation by adding a second order term of the form $h^2 g''(a)/2$; so

$$g(a + h) \approx g(a) + hg'(a) + \frac{h^2}{2} g''(a). \tag{B.3}$$

The right-hand side now matches a further derivative of g at $h = 0$. Generally, a Taylor series expansion for g about the point a with $p+1$ terms has the form

$$g(a + h) \approx g(a) + hg'(a) + \frac{h^2}{2}g''(a) + \cdots + \frac{h^p}{p!}g^{(p)}(a), \qquad (\text{B.4})$$

where $g^{(p)}$ denotes the pth derivative. This pth-degree polynomial matches derivatives of g at the point a from order zero to p.

It is sometimes more convenient to rewrite these expansions with $a + h$ as a single independent variable, say x. The expansions (B.2) and (B.3) then become

$$g(x) \approx g(a) + (x - a)g'(a)$$

and

$$g(x) \approx g(a) + (x - a)g'(a) + \frac{(x - a)^2}{2}g''(a)$$

respectively, and x is required to be close to a for the approximations to be close to g. For the general Taylor expansion (B.4) we then have

$$g(x) \approx g(a) + (x - a)g'(a) + \frac{(x - a)^2}{2}g''(a) + \cdots + \frac{(x - a)^p}{p!}g^{(p)}(a). \qquad (\text{B.5})$$

Figure B.1 illustrates the case of $p = 1, 2, 3$ when $g(x) = x\,\mathrm{e}^{1 - x^2}$ and $a = 1$.

The expansions developed in this chapter are extended in Appendix C to functions of two variables. When the point of expansion is the origin, Taylor series are often referred to as Maclaurin series.

B.2 Accuracy

Increasing the order of our Taylor series expansion by one involves adding a term with an extra power of h. The next term that we would include in the right-hand side of (B.4) to improve the accuracy would be proportional h^{p+1}. Using the asymptotic order notation, as introduced in Section 2.2, we may write

$$g(a + h) = g(a) + hg'(a) + \frac{h^2}{2}g''(a) + \cdots + \frac{h^p}{p!}g^{(p)}(a) + \mathcal{O}(h^{p+1}),$$

as $h \to 0$, to quantify the accuracy in (B.4). For the version in (B.5), this becomes

$$g(x) = g(a) + (x-a)g'(a) + \frac{(x - a)^2}{2}g''(a) + \cdots + \frac{(x - a)^p}{p!}g^{(p)}(a) + \mathcal{O}((x-a)^{p+1}),$$

as $x \to a$.

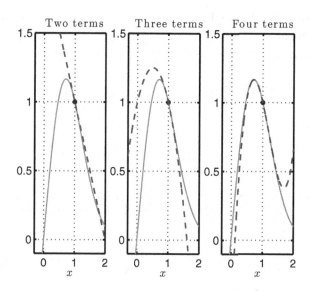

Fig. B.1 In each picture, the function $g(x) = x\,e^{1-x^2}$ is shown as a solid curve. On the left, the Taylor series with two terms (linear) about the point $x = 1$ is shown as a dashed curve. The middle and right pictures show the Taylor series with three terms (quadratic) and four terms (cubic) respectively

The error introduced by the Taylor series can be expressed succinctly, albeit slightly mysteriously. Returning to (B.1), the Mean Value Theorem tells us that the secant approximation $(g(a + h) - g(a))/h$ must match the first derivative of g for at least one point,[1] say c, between a and $a + h$; that is,

$$\frac{g(a + h) - g(a)}{h} = g'(c).$$

This shows that

$$g(a + h) = g(a) + hg'(c). \tag{B.6}$$

This idea extends to a general Taylor series expansion. For (B.4), we have

$$g(a+h) = g(a) + hg'(a) + \frac{h^2}{2}g''(a) + \cdots + \frac{h^p}{p!}g^{(p)}(a) + \frac{h^{p+1}}{(p+1)!}g^{(p+1)}(c), \tag{B.7}$$

where c is a point between a and $a + h$. We emphasize that

---

[1] For a concrete example, suppose that a car is driven at an average speed of 100 miles per hour over a stretch of road. This average speed is computed as the ratio of total distance to total time, corresponding to the secant curve on the left hand side of (B.1). It is intuitively reasonable that this car cannot always have been travelling more slowly than 100 miles per hour over this period, and similarly it cannot always have been travelling more quickly than 100 miles per hour. So at some point in time over that period, it must have been travelling at exactly 100 miles per hour. (Of course more than one such time may exist.)

– although we know that at least one such c exists in that interval, its exact
location is unknown;

– c may be different for each choice of h.

For (B.5), we may write

$$g(x) = g(a) + (x - a)g'(a) + \frac{(x - a)^2}{2}g''(a)+$$

$$\cdots + \frac{(x - a)^p}{p!}g^{(p)}(a) + \frac{(x - a)^{p+1}}{(p + 1)!}g^{(p+1)}(c), \quad \text{(B.8)}$$

where c is a point between a and x.

We refer to (B.7) and (B.8) as Taylor series expansions *with remainder*.

EXERCISES

B.1.* Show that the first four terms in the Taylor series of $g(x) = x\,e^{1-x^2}$
about the point $a = 1$ are

$$g(x) \approx 1 - (x - 1) - (x - 1)^2 + \tfrac{5}{3}(x - 1)^3.$$

B.2.* Determine the first three terms in the Maclaurin expansion of e^{-x}.
Hence, or otherwise, show that the first three non-zero terms in the
Maclaurin expansion of $g(x) = x\,e^{1-x^2}$ are given by

$$g(x) \approx e\big(x - x^3 + \tfrac{1}{2}x^5\big).$$

Jacobians and Variational Equations

The Taylor series expansion of a function of more than one variable was used in the process of linearization in Chapter 12 and, in Section 15.1, it was applied to discrete maps with two components. To see how such expansions can be derived from the scalar version developed in Appendix B, we consider a general map $g : \mathbb{R}^2 \to \mathbb{R}^2$, which we write more explicitly as

$$g(x) = \begin{bmatrix} g_1(x_1, x_2) \\ g_2(x_1, x_2) \end{bmatrix}.$$

Extension to the general case of $g : \mathbb{R}^m \to \mathbb{R}^m$ is straightforward. Now, if we make a small perturbation to the argument x, what happens to the value of g? Introducing a small quantity $\varepsilon \in \mathbb{R}^2$ with components ε_1 and ε_2, we may expand in the first argument—using a two-term Taylor series as described in Appendix B—to obtain

$$g_1(x_1 + \varepsilon_1, x_2 + \varepsilon_2) \approx g_1(x_1, x_2 + \varepsilon_2) + \frac{\partial g_1}{\partial x_1}(x_1, x_2 + \varepsilon_2)\varepsilon_1.$$

Then expanding $g_1(x_1, x_2 + \varepsilon_2)$ in the second argument gives

$$g_1(x_1 + \varepsilon_1, x_2 + \varepsilon_2) \approx g_1(x_1, x_2) + \frac{\partial g_1}{\partial x_2}(x_1, x_2)\varepsilon_2 + \frac{\partial g_1}{\partial x_1}(x_1, x_2 + \varepsilon_2)\varepsilon_1.$$

The final term on the right-hand side could be expanded as $\partial g_1/\partial x_1(x_1, x_2)\varepsilon_1$ plus second order terms in ε_1 and ε_2. So we arrive at the expansion

$$g_1(x_1 + \varepsilon_1, x_2 + \varepsilon_2) \approx g_1(x_1, x_2) + \frac{\partial g_1}{\partial x_1}(x_1, x_2)\varepsilon_1 + \frac{\partial g_1}{\partial x_2}(x_1, x_2)\varepsilon_2. \quad \text{(C.1)}$$

Analogously,

$$g_2(x_1 + \varepsilon_1, x_2 + \varepsilon_2) \approx g_2(x_1, x_2) + \frac{\partial g_2}{\partial x_1}(x_1, x_2)\varepsilon_1 + \frac{\partial g_2}{\partial x_2}(x_1, x_2)\varepsilon_2. \quad \text{(C.2)}$$

The expansions (C.1) and (C.2) can be put together in matrix-vector form as

$$\boldsymbol{g}(\boldsymbol{x} + \boldsymbol{\varepsilon}) \approx \boldsymbol{g}(\boldsymbol{x}) + \frac{\partial \boldsymbol{g}}{\partial \boldsymbol{x}}(\boldsymbol{x})\,\boldsymbol{\varepsilon}, \quad \text{(C.3)}$$

where $\partial \boldsymbol{g}/\partial \boldsymbol{x}$ denotes the *Jacobian* of $\boldsymbol{g}$:

$$\frac{\partial \boldsymbol{g}}{\partial \boldsymbol{x}} = \begin{bmatrix} \dfrac{\partial g_1}{\partial x_1} & \dfrac{\partial g_1}{\partial x_2} \\ \dfrac{\partial g_2}{\partial x_1} & \dfrac{\partial g_2}{\partial x_2} \end{bmatrix}. \quad \text{(C.4)}$$

In Chapter 15 we also looked at the derivatives of an ODE solution with respect to the initial conditions. Here, we briefly explain how the relevant *variational equations* can be derived. We will restrict ourselves to the case of two autonomous ODEs, $\boldsymbol{x}'(t) = \boldsymbol{f}(\boldsymbol{x}(t))$, for $\boldsymbol{x} \in \mathbb{R}^2$, with initial conditions $x_1(0) = a$ and $x_2(0) = b$. We write this more explicitly as

$$\begin{bmatrix} x_1' \\ x_2' \end{bmatrix} = \begin{bmatrix} f_1(x_1, x_2) \\ f_2(x_1, x_2) \end{bmatrix}, \qquad x_1(0) = a, \; x_2(0) = b. \quad \text{(C.5)}$$

Our aim is to derive an ODE for the collection of partial derivatives $\partial x_1/\partial a$, $\partial x_1/\partial b$, $\partial x_2/\partial a$ and $\partial x_2/\partial b$. Differentiating $\partial x_1/\partial a$ with respect to time, interchanging the order of the derivatives, and using the chain rule, we find that

$$\frac{\mathrm{d}}{\mathrm{d}t}\left(\frac{\partial x_1}{\partial a}\right) = \frac{\partial}{\partial a}\left(\frac{\mathrm{d}x_1}{\mathrm{d}t}\right) = \frac{\partial}{\partial a}\left(f_1(x_1, x_2)\right) = \frac{\partial f_1}{\partial x_1}\frac{\partial x_1}{\partial a} + \frac{\partial f_1}{\partial x_2}\frac{\partial x_2}{\partial a}.$$

Analogously, we have

$$\frac{\mathrm{d}}{\mathrm{d}t}\left(\frac{\partial x_1}{\partial b}\right) = \frac{\partial f_1}{\partial x_1}\frac{\partial x_1}{\partial b} + \frac{\partial f_1}{\partial x_2}\frac{\partial x_2}{\partial b},$$

$$\frac{\mathrm{d}}{\mathrm{d}t}\left(\frac{\partial x_2}{\partial a}\right) = \frac{\partial f_2}{\partial x_1}\frac{\partial x_1}{\partial a} + \frac{\partial f_2}{\partial x_2}\frac{\partial x_2}{\partial a},$$

$$\frac{\mathrm{d}}{\mathrm{d}t}\left(\frac{\partial x_2}{\partial b}\right) = \frac{\partial f_2}{\partial x_1}\frac{\partial x_1}{\partial b} + \frac{\partial f_2}{\partial x_2}\frac{\partial x_2}{\partial b}.$$

Putting the four partial derivatives into a matrix, these relations may be written

$$\frac{\mathrm{d}}{\mathrm{d}t} \begin{bmatrix} \dfrac{\partial x_1}{\partial a} & \dfrac{\partial x_1}{\partial b} \\ \dfrac{\partial x_2}{\partial a} & \dfrac{\partial x_2}{\partial b} \end{bmatrix} = \begin{bmatrix} \dfrac{\partial f_1}{\partial x_1} & \dfrac{\partial f_1}{\partial x_2} \\ \dfrac{\partial f_2}{\partial x_1} & \dfrac{\partial f_2}{\partial x_2} \end{bmatrix} \begin{bmatrix} \dfrac{\partial x_1}{\partial a} & \dfrac{\partial x_1}{\partial b} \\ \dfrac{\partial x_2}{\partial a} & \dfrac{\partial x_2}{\partial b} \end{bmatrix}. \quad \text{(C.6)}$$

Here, the matrix

$$\begin{bmatrix} \dfrac{\partial f_1}{\partial x_1} & \dfrac{\partial f_1}{\partial x_2} \\ \dfrac{\partial f_2}{\partial x_1} & \dfrac{\partial f_2}{\partial x_2} \end{bmatrix}, \qquad (C.7)$$

which is known as the *Jacobian* of the ODE system, is to be evaluated along the solution $x_1(t)$, $x_2(t)$ of (C.5).

D
Constant-Coefficient Difference Equations

We will describe here means of solving simple linear constant-coefficient difference equations ($\triangle$Es for short). For a more in-depth introduction we recommend the books by Dahlquist and Björk [17, Section 3.3.5] or Elaydi [18].

A kth order difference is a relationship between $k+1$ consecutive terms of a sequence $x_0, x_1, \ldots, x_n, \ldots$. In a kth-order linear constant-coefficient $\triangle$E this relationship is of the form

$$a_k x_{n+k} + a_{k-1} x_{n+k-1} + \cdots + a_0 x_n = f_n \tag{D.1}$$

in which the coefficients are $a_k, a_{k-1}, \ldots, a_0$, and f_n is a given sequence of numbers. We shall assume throughout that n runs consecutively through the non-negative integers: $n = 0, 1, \ldots$. We shall also assume that neither a_0 nor a_k is zero, for otherwise this could be written as a $\triangle$E of order lower than k.

Our objective is: given $k > 0$, a set of coefficients and the sequence f_n to obtain a formula for the nth term of the sequence satisfying (D.1). We shall focus mainly on first ($k = 1$) and second-order ($k = 2$) $\triangle$Es and also consider only the cases when either $f_n \equiv 0$ for all n (called the homogeneous case) or when the forcing term f_n has a particularly simple form.

As for linear constant-coefficient ODEs, the general solution of $\triangle$Es of the form (D.1) may be composed of the sum of a *complementary function* (CF—the general solution of the homogeneous $\triangle$E) and a particular solution (PS—any solution of the given $\triangle$E). The arbitrary constants in a general solution may be fixed by specifying the appropriate number (k) of starting conditions: $x_j = \eta_j$, for $j = 0 : k - 1$.

D.1 First-order Equations

Consider the first-order linear constant-coefficient problem:

$$x_{n+1} = ax_n + b, \qquad n = 0, 1, \ldots. \tag{D.2}$$

By setting, in turn, $n = 0, 1, 2$ we can compute

$$
\begin{aligned}
x_1 &= ax_0 + b, \\
x_2 &= ax_1 + b = a^2 x_0 + (a+1)b, \\
x_3 &= ax_2 + b = a^3 x_0 + (a^2 + a + 1)b
\end{aligned}
$$

and, continuing in the same vein, we can clearly compute any term in the sequence x_n, although the nature of the nth term may not be obvious.

A systematic approach first considers the homogeneous equation $x_{n+1} = ax_n$ for which $x_1 = ax_0$, $x_2 = ax_1 = a^2 x_0$ and, by induction, the nth term is $x_n = a^n x_0$, for any value of x_0. Setting $x_0 = A$, an arbitrary constant, the CF is given by

$$x_n = Aa^n.$$

We next look for a particular solution. Since the right-hand side of (D.2) is constant, we look for a PS in the form of a constant sequence $x_n = C$. Substituting $x_n = x_{n+1} = C$ into (D.2) leads to

$$C = \frac{b}{1-a}, \qquad (a \neq 1).$$

Constant solutions are often referred to as *fixed points* (FPs) of the $\triangle$E (see Chapter 12—Long-Term Dynamics).

The sum of the CF and PS:

$$x_n = Aa^n + \frac{b}{1-a} \qquad (a \neq 1), \tag{D.3}$$

is seen, by substitution, to be a solution of (D.2). It can be proved that there are no other solutions of this equation and that this is, therefore, the general solution (GS) provided that $a \neq 1$.

When $a = 1$ the recurrence becomes

$$x_{n+1} - x_n = b,$$

so, by the telescoping series property,

$$
\begin{aligned}
x_n &= (x_n - x_{n-1}) + (x_{n-1} - x_{n-2}) + \cdots + (x_1 - x_0) + x_0 \\
&= b + b + \cdots + b + x_0 \\
&= nb + x_0.
\end{aligned}
$$

Hence, the GS in this case is

$$x_n = A + nb,$$

where A is an arbitrary constant.

Before proceeding to the solution of second-order equations, we note that the GS of (D.2) allows us to find the GS of the $\triangle$E

$$x_{n+1} = ax_n + bk^n, \qquad n = 0, 1, \ldots \qquad \text{(D.4)}$$

by means of the substitution (known as the variation of constants) $x_n = k^n y_n$. We find that y_n satisfies

$$y_{n+1} = \left(\frac{a}{k}\right) y_n + \frac{b}{k}, \qquad \text{(D.5)}$$

which is of the form (D.2). By comparing with the solution of (D.2) it is readily shown that (D.5) has GS

$$y_n = \begin{cases} A(a/k)^n + \dfrac{b/k}{1 - a/k} & (a/k \neq 1), \\ A + (b/k)n & (a/k = 1). \end{cases}$$

Hence, using $x_n = k^n y_n$, it is seen that (D.4) has GS

$$x_n = \begin{cases} Aa^n + \dfrac{b}{k-a} k^n & (a \neq k), \\ Ak^n + bnk^{n-1} & (a = k). \end{cases} \qquad \text{(D.6)}$$

D.2 Second-order Equations

We begin by looking for the GS of the homogeneous equation

$$x_{n+2} + ax_{n+1} + bx_n = 0. \qquad \text{(D.7)}$$

This equation will have solutions of the form $x_n = Ar^n$, where A is an arbitrary constant, provided that r is a root of the quadratic equation

$$r^2 + ar + b = 0, \qquad \text{(D.8)}$$

known as the *auxiliary equation*. Suppose that this equation has roots α and β, then, since $(r - \alpha)(r - \beta) = r^2 - (\alpha + \beta)r + \alpha\beta$, it follows that

$$a = -(\alpha + \beta), \qquad b = \alpha\beta.$$

Thus, the inhomogeneous $\triangle$E

$$x_{n+2} + ax_{n+1} + bx_n = f_n \qquad \text{(D.9)}$$

may be written as $x_{n+2} - (\alpha + \beta)x_{n+1} + \alpha\beta x_n = f_n$. That is,

$$\left(x_{n+2} - \alpha x_{n+1}\right) - \beta\left(x_{n+1} - \alpha x_n\right) = f_n.$$

By defining $y_n = x_{n+1} - \alpha x_n$ this becomes $y_{n+1} - \beta y_n = f_n$ and we have succeeded in rewriting (D.9) as the first-order system of $\triangle$Es

$$\left.\begin{array}{c} x_{n+1} - \alpha x_n = y_n \\ y_{n+1} - \beta y_n = f_n \end{array}\right\}. \tag{D.10}$$

In the homogeneous case $f_n \equiv 0$ the second of these has GS $y_n = C\beta^n$, where C is an arbitrary constant, and then the first becomes

$$x_{n+1} - \alpha x_n = C\beta^n.$$

This has the same form as (D.4) and so we can deduce its GS immediately

$$x_n = \begin{cases} A\alpha^n + B\beta^n & (\alpha \neq \beta), \\ (A + Bn)\alpha^n & (\alpha = \beta), \end{cases} \tag{D.11}$$

where A and B are arbitrary constants (B can be expressed in terms of α, β and C).

Notice that, in the case that $\alpha = \beta$ and $|\alpha| = 1$, then $|x_n| \to \infty$ in general as $n \to \infty$, and this is the reason that multiple roots lying on the unit circle in the complex plane had to be excluded in the definition of the root condition used for zero-stability (Definitions 5.4 and 5.5).

The only inhomogeneous case of (D.9) that we will consider is for a constant forcing term, in which case $f_n \equiv f$ for all n. Then, the second of (D.10) has GS

$$y_n = \begin{cases} C\beta^n + \dfrac{f}{1 - \beta} & (\beta \neq 1), \\ C + nf & (\beta = 1). \end{cases}$$

When this is substituted into the right-hand side of the first equation in (D.10) we can employ a similar process to that in the previous section to find[1]

$$x_n = \begin{cases} A\alpha^n + B\beta^n + \dfrac{f}{(1 - \alpha)(1 - \beta)}, & (\alpha \neq 1, \beta \neq 1, \alpha \neq \beta) \\ (A + Bn)\alpha^n + \dfrac{f}{(1 - \alpha)^2} & (\alpha = \beta \neq 1), \\ A\alpha^n + B + \dfrac{f}{1 - \alpha}n & (\alpha \neq \beta = 1), \\ An + B\beta^n + \dfrac{f}{1 - \beta}n & (\beta \neq \alpha = 1), \\ A + Bn + \frac{1}{2}fn^2, & (\alpha = \beta = 1). \end{cases}$$

---

[1]It may be verified by substitution that the last of these results, $x_n = A + Bn + \frac{1}{2}fn^2$, satisfies $x_{n+2} - 2x_{n+1} + x_n = f$.

D.3 Higher Order Equations

Seeking solutions of the homogeneous equation

$$a_k x_{n+k} + a_{k-1} x_{n+k-1} + \cdots + a_0 x_n = 0$$

in the form $x_n = Ar^n$ requires r to be a root of the auxiliary equation

$$a_k r^k + a_{k-1} r^{k-1} + \cdots + a_0 = 0. \tag{D.12}$$

For each root r of this equation, suppose that it has multiplicity m, then there is a contribution

$$P(n)r^n$$

to the CF, where $P(n)$ is a polynomial in n of degree $m-1$ (and so contains m arbitrary coefficients). For example, when $m = 1$, the polynomial $P(n)$ has degree 0 and is, therefore, a constant $(P(n) = A)$; when $m = 2$, the polynomial $P(n)$ has degree 1 and has the form $P(n) = A + Bn$. The full CF is the sum of all such terms over all roots of the auxiliary equation.

EXERCISES

D.1.* Determine general solutions of the $\triangle$Es

(a) $2x_{n+1} = x_n + 3$, (b) $x_{n+1} = 2x_n + 3$, (c) $x_{n+1} = x_n + 3$.

Find also the solutions satisfying the initial condition $x_0 = 5$ in each case. Discuss the behaviour of solutions as $n \to \infty$.

D.2.** Use Equation (D.3) to show that the solution of (D.2) that satisfies $x_0 = \eta$ is given by

$$x_n = \eta a^n + b \frac{1 - a^n}{1 - a} \quad (a \neq 1). \tag{D.13}$$

By employing l'Hôpital's rule, show that this reduces to $x_n = \eta + nb$ in the limit $a \to 1$.

Verify this result by showing that (D.13) may be written

$$x_n = \eta a^n + b(1 + a + a^2 + \cdots + a^{n-1})$$

and then setting $a = 1$.

D.3.** Determine general solutions of the $\triangle$Es

a) $2x_{n+1} = x_n + 3 \times 2^n$,

b) $x_{n+1} = 2x_n + 3 \times 2^n$,

c) $x_{n+1} = x_n + 3 \times 2^n$.

D.4.** Determine general solutions of the $\triangle$Es

(a) $x_{n+2} - 3x_{n+1} - 4x_n = 3$, (b) $x_{n+2} - 3x_{n+1} - 4x_n = 3 \times 2^n$,
(c) $x_{n+2} - 4x_{n+1} + 4x_n = 3$, (d) $x_{n+2} - 4x_{n+1} + 4x_n = 3 \times 2^n$.

In case (a), find the solution satisfying the starting conditions $x_0 = 0$ and $x_1 = -1$.

D.5.** Show that the auxiliary equation of the $\triangle$E

$$x_{n+5} - 8x_{n+4} + 25x_{n+3} - 38x_{n+2} + 28x_{n+1} - 8x_n = 0$$

can be factorized as $(r - 2)^3 (r - 1)^2 = 0$. Hence, find the GS of the $\triangle$E.

Bibliography

[1] U. Alon. *An Introduction to Systems Biology.* Chapman & Hall/CRC, London, 2006.

[2] U. M. Ascher and L. R. Petzold. *Computer Methods for Ordinary Differential Equations and Differential-Algebraic Equations.* SIAM, Philadelphia, USA, 1998.

[3] W.-J. Beyn. Numerical methods for dynamical systems. In W. Light, editor, *Advances in Numerical Analysis, Vol. 1*, pages 175–236. Clarendon Press, 1991.

[4] P. Bogacki and L. F. Shampine. A 3(2) pair of Runge–Kutta formulas. *Appl. Math. Lett.*, 2:1–9, 1989.

[5] M. Braun. *Differential Equations and their Applications.* Springer-Verlag, 1983.

[6] J. C. Butcher. *Numerical Methods for Ordinary Differential Equations.* Wiley, Chichester, UK, 2nd edition, 2008.

[7] B. Calderhead, M. Girolami, and D. J. Higham. Is it safe to go out yet? Statistical inference in a zombie outbreak model. Technical Report 6, University of Strathclyde, Glasgow, 2010. To appear in R. J. Smith?, editor, *"Mathematical Modelling of Zombies"*, University of Ottawa Press.

[8] S. Chow and E. V. Vleck. A shadowing lemma approach to global error analysis for initial value ODEs. *SIAM J. Sci. Comput.*, 15:959–976, 1994.

[9] L. Cobb. Stochastic differential equations for the Social Sciences. In L. Cobb and R. M. Thrall, editors, *Mathematical Frontiers of the Social and Policy Sciences.* Westview Press, 1981.

[10] G. J. Cooper. Stability of Runge–Kutta methods for trajectory problems. *IMA J. Numer. Anal.*, 7:1–13, 1987.

[11] J. C. Cox, J. E. Ingersoll, and S. A. Ross. A theory of the term structure of interest rates. *Econometrica*, 53:385–407, 1985.

[12] A. Croft and R. Davidson. *Mathematics for Engineers*. Pearson/Prentice Hall, Harlow, UK, 3rd edition, 2008.

[13] S. Cyganowski, P. Kloeden, and J. Ombach. *From Elementary Probability to Stochastic Differential Equations with MAPLE*. Springer, Berlin, 2002.

[14] G. Dahlquist. Convergence and stability in the numerical integration of ordinary differential equations. *Math. Scand.*, 4:33–53, 1956.

[15] G. Dahlquist. Stability and error bounds in the numerical integration of ordinary differential equations. *Trans. of the R. Inst. Technol. Stockholm, Sweden*, (130):87pp, 1959.

[16] G. Dahlquist and A. Björk. *Numerical Methods*. Prentice Hall, 1974.

[17] G. Dahlquist and A. Björk. *Numerical Methods in Scientific Computation*. SIAM, Philadelphia, USA, 2008.

[18] S. Elaydi. *An Introduction to Difference Equations*. Undergraduate Texts in Mathematics. Springer-Verlag, 3rd edition, 2005.

[19] K. Eriksson, D. Estep, P. Hansbo, and C. Johnson. *Computational Differential Equations*. Cambridge University Press, 1996.

[20] G. Fulford, P. Forrester, and A. Jones. *Modelling with Differential and Difference Equations*, volume 10 of *Australian Mathematical Society Lecture Series*. Cambridge University Press, 1997.

[21] M. B. Giles. Multilevel Monte Carlo path simulation. *Operations Research*, 56:607–617, 2008.

[22] O. Gonzalez, D. J. Higham, and A. M. Stuart. Qualitative properties of modified equations. *IMA J. Numer. Anal.*, 19:169–190, 1999.

[23] A. Griewank. *Evaluating Derivatives: Principles and Techniques of Algorithmic Differentiation*, volume 19 of *Frontiers in Applied Mathematics*. SIAM, 2000.

[24] D. F. Griffiths and J. M. Sanz-Serna. On the scope of the method of modified equations. *SIAM J. Sci. Stat. Comput.*, 7:994–1008, 1986.

[25] D. F. Griffiths, P. K. Sweby, and H. C. Yee. On spurious asymptotic numerical solutions of explicit Runge–Kutta methods. *IMA J. Num. Anal.*, 12:319–338, 1992.

[26] E. Hairer, C. Lubich, and G. Wanner. *Geometric Numerical Integration: Structure-Preserving Algorithms for Ordinary Differential Equations.* Springer, Berlin, 2002.

[27] E. Hairer, C. Lubich, and G. Wanner. Geometric numerical integration illustrated by the Störmer–Verlet method. In A. Iserles, editor, *Acta Numerica*, pages 399–450. Cambridge University Press, 2003.

[28] E. Hairer, S. P. Nørsett, and G. Wanner. *Solving Ordinary Differential Equations I: Nonstiff Problems.* Springer-Verlag, Berlin, 2nd revised edition, 1993.

[29] E. Hairer and G. Wanner. *Solving Ordinary Differential Equations II: Stiff and Differential-Algebraic Problems.* Springer-Verlag, Berlin, 2nd edition, 1996.

[30] J. K. Hale and H. Kocak. *Dynamics and Bifurcations.* Springer, Berlin, 3rd edition, 1991.

[31] D. J. Higham. An algorithmic introduction to numerical simulation of stochastic differential equations. *SIAM Review*, 43:525–546.

[32] D. J. Higham. *An Introduction to Financial Option Valuation: Mathematics, Stochastics and Computation.* Cambridge University Press, Cambridge, 2004.

[33] D. J. Higham. Modeling and simulating chemical reactions. *SIAM Rev.*, 50:347–368, 2008.

[34] D. J. Higham and N. J. Higham. MATLAB *Guide.* SIAM, Philadelphia, USA, 2nd edition, 2005.

[35] D. J. Higham and L. N. Trefethen. Stiffness of ODEs. *BIT Numer. Math.*, 33:285–303, 1993.

[36] N. J. Higham. *Accuracy and Stability of Numerical Algorithms.* SIAM, Philadelphia, 1996.

[37] N. J. Higham. *Functions of Matrices: Theory and Computation.* SIAM, Philadelphia, PA, USA, 2008.

[38] K. E. Hirst. *Calculus of One Variable.* Springer (Springer Undergraduate Mathematics Series), Berlin, 2006.

[39] A. Iserles. *A First Course in the Numerical Analysis of Differential Equations.* Cambridge University Press, 2nd edition, 2008.

[40] E. I. Jury. *Theory and Application of the z-Transform Method.* John Wiley and Sons, 1964.

[41] C. T. Kelley. *Iterative Methods for Linear and Nonlinear Equations*, volume 16 of *Frontiers in Applied Mathematics*. SIAM, Philadelphia, USA, 1995.

[42] P. E. Kloeden and E. Platen. *Numerical Solution of Stochastic Differential Equations*. Springer Verlag, Berlin, third printing, 1999.

[43] Y. K. Kwok. *Mathematical Models of Financial Derivatives*. Springer, Berlin, 1998.

[44] J. D. Lambert. *Numerical Methods in Ordinary Differential Systems*. John Wiley and Sons, Chichester, 1991.

[45] B. Leimkuhler and S. Reich. *Simulating Hamiltonian Dynamics*. Cambridge University Press, 2004.

[46] D. Lesmono, E. J. Tonkes, and K. Burrage. An early political election problem. In J. Crawford and A. J. Roberts, editors, *Proceedings of 11th Computational Techniques and Applications Conference CTAC-2003*, volume 45, pages C16–C33, 2003.

[47] R. J. Leveque. *Finite Difference Methods for Ordinary and Partial Differential Equation*. SIAM, Philadelphia, USA, 2007.

[48] D. J. Logan. *A First Course in Differential Equations*. Undergraduate Texts in Mathematics. Springer-Verlag, 2006.

[49] X. Mao. *Stochastic Differential Equations and Applications*. Horwood, Chichester, second edition, 2008.

[50] R. I. McLachlan and G. R. W. Quispel. Geometric integrators for ODEs. *J. Phys. A*, 39:5251–5286, 2006.

[51] T. Mikosch. *Elementary Stochastic Calculus (with Finance in View)*. World Scientific, Singapore, 1998.

[52] G. N. Milstein and M. V. Tretyakov. *Stochastic Numerics for Mathematical Physics*. Springer-Verlag, Berlin, 2004.

[53] C. B. Moler. *Numerical Computing with* MATLAB. SIAM, 2004.

[54] K. W. Morton and D. F. Mayers. *Numerical Solution of Partial Differential Equations*. Cambridge University Press, 2nd edition, 2005.

[55] P. Munz, I. Hudea, J. Imad, and R. J. Smith?. When zombies attack!: Mathematical modelling of an outbreak of zombie infection. In J. Tchuenche and C. Chiyaka, editors, *Infectious Disease Modelling Research Progress*, pages 133–150, Nova, 2009.

[56] J. D. Murray. *Mathematical Biology*, volume 17 of *Interdisciplinary Applied Mathematics*. Springer-Verlag, 3rd edition, 2002.

[57] R. K. Nagle, E. B. Saff, and A. D. Snider. *Fundamentals of Differential Equations and Boundary Value Problems*. Pearson Education Inc., 4th edition, 2004.

[58] P. J. Nahin. *Chases and Escapes: The Mathematics of Pursuit and Evasion*. Princeton University Press, Princeton, USA, 2007.

[59] R. E. O'Malley, Jr. *Thinking About Ordinary Differential Equations*, volume 18 of *Cambridge Texts in Applied Mathematics*. Cambridge University Press, 1997.

[60] J. M. Sanz-Serna. Geometric integration. In I. S. Duff and G. A. Watson, editors, *The State of the Art in Numerical Analysis*, pages 121–143. Oxford University Press, 1997.

[61] J. M. Sanz-Serna and M. P. Calvo. *Numerical Hamiltonian Problems*, volume 7 of *Applied Mathematics and Mathematical Computation*. Chapman & Hall, 1994.

[62] L. F. Shampine. *Numerical Solution of Ordinary Differential Equations*. Chapman & Hall, New York, USA, 1994.

[63] L. F. Shampine, I. Gladwell, and S. Thompson. *Solving ODEs with MATLAB*. Cambridge University Press, 2003. ISBN: 0-521-53094-6.

[64] J. Stewart. *Calculus*. Thomson Brooks/Cole, Belmont, USA, 6th edition, 2009.

[65] A. M. Stuart and A. R. Humphries. *Dynamical Systems and Numerical Analysis*. Cambridge University Press, Cambridge, 1996.

[66] J. M. T. Thompson and H. B. Stewart. *Nonlinear Dynamics and Chaos, Geometrical Methods for Engineers and Scientists*. John Wiley and Sons, 2nd edition, 2002.

[67] L. N. Trefethen and M. Embree. *Spectra and Pseudospectra: The Behavior of Nonnormal Matrices and Operators*. Princeton University Press, Princeton, 2005.

[68] F. Verhulst. *Nonlinear Differential Equations and Dynamical Systems*. Universitext. Springer-Verlag, 2nd edition, 1990.

[69] D. J. Wilkinson. *Stochastic Modelling for Systems Biology*. Chapman & Hall/CRC, 2006.

[70] F. C. Wilson and S. Adamson. *Applied Calculus*. Houghton Mifflin Harcourt, Boston, USA, 2009.

Index